INTERFACIAL SYNTHESIS

INTERFACIAL SYNTHESIS

(IN TWO VOLUMES)

VOLUME I
Fundamentals

Edited by

FRANK MILLICH
*Polymer Chemistry Division
Department of Chemistry
University of Missouri—Kansas City
Kansas City, Missouri*

CHARLES E. CARRAHER, Jr.
*Department of Chemistry
Wright State University
Dayton, Ohio*

MARCEL DEKKER, INC.   New York and Basel

Library of Congress Cataloging in Publication Data
Main entry under title:

Interfacial synthesis.

   Includes indexes.
   CONTENTS: v. 1. Fundamentals.
   1. Polymers and polymerization. 2. Chemistry,
Organic--Synthesis. I. Millich, Frank. II. Carraher,
Charles E.
QD281.P6I48   547'.28   75-27750
ISBN 0-8247-6372-6

COPYRIGHT © 1977 by MARCEL DEKKER, INC. ALL RIGHTS RESERVED.

Neither this book nor any part may be reproduced or transmitted, in any form or by any means, electronic or mechanical, including photocopying, microfilming, and recording, or by any information storage and retrieval system, without permission in writing from the publisher.

MARCEL DEKKER, INC.
270 Madison Avenue, New York, New York 10016

Current printing (last digit):
10 9 8 7 6 5 4 3 2 1

PRINTED IN THE UNITED STATES OF AMERICA

This book is dedicated with affection in memory of my father, Frank J., and to my mother, Frances, for her many sacrifices, love, and the devotion that was such an important part of my youth, and to my children, Theadocia and Frank "Teddy", who make my labors worthwhile.

                                        Frank Millich

And, to my wife Loyalea and the children, Charles III, Shawn, Michelle, Erin, Heather, and Colleen.

                                  Charles E. Carraher, Jr.

PREFACE

This two-volume book deals with the present-day exercise of interfacial synthesis, a technology that has not received much attention or exploitation, aside from limited application to the preparation of some linear condensation polymers. Volume I (Chapters 1 through 10) is devoted to fundamentals; Volume II (Chapters 11 through 23) discusses polymer application and technology.

The first book in this field appeared in 1965: *Condensation Polymers: By Interfacial and Solution Methods,* authored by Paul W. Morgan. This hallmark compilation of experiences with polycondensation contains much that is relevant today. Therefore, in Volume II we include only material which is supplementary to Morgan's text, to avoid unnecessary repetition. Also included are some new polymer classes.

This text differs from its predecessor in its emphasis on selected aspects of interfacial synthesis. The technique represents that of a facile laboratory procedure which has not achieved routine introduction into college laboratories. By and large, the present generation of chemistry students and faculty are unfamiliar with details of the process; at the same time, the mechanisms of reactions in heterophasic systems are not thoroughly defined. We are only beginning to understand the mechanisms and the methods for experimental study of interfacial polymerization. In an effort to stimulate further study, Volume I was assembled with students of chemistry in mind. It is introductory and is intended to lead the interested reader to the chemical literature. In a sense, then, this book has begun a transition toward a more fundamental treatment of the subject—a trend, which it is hoped, will progress further with succeeding volumes.

Contributions were invited only from experts who have firsthand experience and who have published on their respective subjects. Half the authors are from foreign countries. Special benefits which they bring are summaries of foreign literature sources that are not convenient or always accessible to everyone, and, on occasion, defense of refreshing viewpoints not always shared by English-speaking scholars. At times, the editors have had to transliterate grammatical expressions from foreign contributors. In so doing, we have made minimal changes in order to stay as close as possible to the author's intent. In such circumstances the styles of the various contributors show unavoidable variations. The authors have been given the reins in their treatment of the subjects, which they are best to judge. In almost all cases, each chapter has been reviewed by at least two competent reviewers in addition to review and re-review by the editors, in the hope of achieving the maximum value with each contribution; the authors reserved the right, of course, to accept or reject suggested changes.

The editors sincerely hope that they have served their fellow scientists well. We wish to express our gratitude to our many colleagues who graciously gave advice or assisted with chapter reviews. We especially wish to thank Drs. Ian J. Alexander, Robert Boschan, Stephen D. Bruck, Howard G. Clark, Robert J. Cotter, William M. Eareckson, John Eichelberger, David R. Hackathorn, Robert F. Henry, D. B. G. Jacquiss, John E. McMurry, Leszek Makaruk, Frederic M. Menger, Norman E. Miller, Delbert D. Reynolds, John R. Schaefgen, George P. Scott, William L. Wasley, Robert E. Whitfield, Emerson L. Wittbecker, and many of the contributing authors who occasionally were asked to review another's manuscript. We also thank Charles E. Carraher III for his help with the indexing. A large debt of gratitude is owed Dr. Paul W. Morgan of E. I. du Pont de Nemours and Co., Wilmington, Delaware, who has been consulted on many occasions, dating from the first inception of this book.

CONTENTS

	PREFACE	v
	CONTRIBUTORS TO VOLUME I	ix
	CONTENTS OF VOLUME II	xi
1.	INTRODUCTION, *Frank Millich*	1
2.	STIRRING IN ORGANIC CHEMICAL SYNTHESIS, *J. Henry Rushton*	9
3.	HIGH-SPEED STIRRING IN INTERFACIAL SYNTHESIS, *J. Henry Rushton*	37
4.	APPLICATION OF TWO-PHASE SYNTHESIS IN ORGANIC CHEMISTRY, *David J. Goldsmith*	57
5.	KINETICS AND MECHANISMS, *J. Howard Bradbury and Peter J. Crawford*	77
6.	INTERFACE EFFECTS ON CHEMICAL REACTION RATE, *F. MacRitchie*	103
7.	LIQUID-VAPOR INTERFACIAL POLYCONDENSATION, *L. B. Sokolov*	141
8.	COPOLYCONDENSATION AND MACROSCOPIC KINETICS, *L. B. Sokolov and V. Z. Nikonov*	167
9.	INTERFACIAL, COLLOIDAL, AND KINETIC ASPECTS OF EMULSION POLYMERIZATION, *John L. Gardon*	205
10.	BIOLOGICAL PHENOMENA AND INTERFACES, *David Allan Cadenhead and Robert E. Baier*	255
	AUTHOR INDEX	281
	SUBJECT INDEX	291

CONTRIBUTORS TO VOLUME I

ROBERT E. BAIER, *Department of Biophysics, Roswell Park Division, State University of New York at Buffalo, New York*

J. HOWARD BRADBURY, *Chemistry Department, Australian National University, Canberra, Australia*

DAVID ALLAN CADENHEAD, *Department of Chemistry, State University of New York at Buffalo, New York*

PETER J. CRAWFORD,* *Commonwealth Scientific and Industrial Research Organisation, Campbell, Australia*

JOHN L. GARDON, *Research Department, Coatings and Ink Division, M & T Chemicals, Inc., Southfield, Michigan*

DAVID J. GOLDSMITH, *Department of Chemistry, Emory University, Atlanta, Georgia*

F. MACRITCHIE, *C.S.I.R.O. Wheat and Research Unit, North Ryde, Australia*

FRANK MILLICH, *Department of Chemistry, Polymer Chemistry Division, University of Missouri--Kansas City, Missouri*

V. Z. NIKONOV, *Department of Polycondensation Processes, All-Union Research Institute of Synthetic Resins, Vladimir, USSR*

J. HENRY RUSHTON, *School of Chemical Engineering, Purdue University, Lafayette, Indiana; also Mixing Equipment Co., Rochester, New York*

L. B. SOKOLOV, *Department of Polycondensation Processes, All-Union Research Institute of Synthetic Resins, Vladimir, USSR*

---

*Present affiliation: *Department of Environment, Housing, and Community Development, Canberra, Australia.*

CONTENTS OF VOLUME II

11. COMMERCIAL APPLICATIONS OF INTERFACIAL SYNTHESIS, *Earl D. Oliver and Yen-Chen Yen*

12. POLYCARBOXYLIC ESTERS, *Samuel C. Temin*

13. POLYCARBONATES, *Hugo Vernaleken*

14. PREPARATION OF POLYAMIDES BY INTERFACIAL POLYADDITION OF CARBON SUBOXIDE TO DIAMINES, *Irena Daniewska*

15. POLYAMIDES, *V. Z. Nikonov and V. M. Savinov*

16. POLYESTERAMIDES, *Ivan M. Panayotov*

17. POLYURETHANES, *Takehide Tanaka and Tetsuo Yokoyama*

18. THE INTERFACIAL PREPARATION OF POLYUREAS, *Kenneth C. Stueben and Austin E. Barnabeo*

19. POLYPHOSPHONATES, POLYPHOSPHATES, AND POLYPHOSPHITES, *Frank Millich, Larry L. Lambing, and J. Teague*

20. OTHER PHOSPHORUS-CONTAINING POLYMERS, *Charles E. Carraher, Jr.*

21. ORGANOMETALLIC POLYMERS, *Charles E. Carraher, Jr.*

22. MODIFICATION OF NATURAL POLYMERS BY INTERFACIAL METHODS, *M. Horio*

23. INTERFACIAL MODIFICATION OF POLY(VINYL ALCOHOL) AND RELATED POLYMERS, *Minoru Tsuda*

AUTHOR INDEX

SUBJECT INDEX

1.

INTRODUCTION

Frank Millich

Polymer Chemistry Division
Department of Chemistry
University of Missouri--Kansas City
Kansas City, Missouri

The connotation of the term *interfacial synthesis* implies chemical formation at the interfaces of two or more distinct phases. Many organic chemists have had the experience of conducting a Schotten-Baumann synthesis of an amide and the hydrogenation of unsaturated compounds by means of hydrogen gas, a solvent, and a solid catalyst. Many bioorganic chemists may have conducted a reaction proceeding on the surface of a membrane. Many an educator may have demonstrated the "nylon rope trick" in his lecture class. And many a polymer chemist may have formed a condensation polymer in a laboratory high-speed blender. Unfortunately for a number of these experimentalists such experiences may represent the total depth of involvement in the intricacies of interfacial synthesis. Except for some very significant islands of intensive interests such as heterophasic catalysis, emulsion free radical polymerization (c.f. Chap. 9), interfacial polycondensation, thin-film and membrane phenomena, and chemical evolution of life, in-depth study of interfacial synthesis as a coordinated integrated field of chemical knowledge is largely undeveloped. Biochemical research especially illustrates the large unbridged gap that exists between the bioorganic chemist, standing on one shore of his approach to interfacial systems, and the polymer chemist, armed with more conventional disciplinary tools and interests, standing on the other. One can appreciate the potential breadth of the field connoted by the term "interfacial synthesis"; one can also appreciate that, as a broad integrated field, it is still in its infancy, and that, pregnant with significance, it will be a vital and rewarding field for future cultivation.

By activities and broad training the polymer chemist is perhaps most centrally located among the islands of specialization mentioned above, and much valuable data on interfacial polycondensation has been accumulated and summarized in Dr. Paul W. Morgan's 1965 text, *Condensation Polymers: By Interfacial and Solution Methods*. Volume II of this book is devoted to a continuation of Morgan's ledger of experience. The editors' interest in interfacial synthesis of polyorganophosphates, as nucleic acid analogs, moved them to organize a book that would supplement Morgan's text. From the beginning it was assumed that this anthology would continue with periodic extension beyond the present book. Thus Volume II is not intended to be comprehensive, although some of the authors have invested considerable effort in tabular surveys of polymers reported to have been made by interfacial methods. When reporting on a class of polymers that had prior coverage in Morgan's book, the authors concentrated on aspects of the subject germane to the decade since Morgan's text was published. Furthermore, chapters appear in this book that are not covered at any length in Morgan's text. Thus the reader will find treatment of organometallic (Chap. 21) and various phosphorus-containing polymer syntheses (Chaps. 19 and 20). He will find some emphasis on polymer modification by interfacial synthesis, involving natural polymers (Chap. 22) and synthetic vinyl polymers (Chap. 23). He will also read about the problems encountered when, in the interest of achieving greater synthetic scope, interfacial synthesis is extended to copolycondensations (Chap. 8).

It is possible to compare the results of influential reaction parameters in the synthesis of a given polymer produced by alternative techniques. High- and low-temperature solution condensations

1. Introduction

may be contrasted; the latter may be contrasted with the reaction of unstirred layered solutions; and these may in turn be contrasted with stirred and unstirred interfacial polycondensation reactions. Even the gaseous state contrasts with the dissolved state of a reagent in its effect on condensation (Chap. 7).

Most graduate chemistry students are required to take one or more courses on kinetics of chemical reactions. However, their teachers are often quite silent on the subject of interfaces and heterophasic reaction systems. Reaction mechanisms and kinetic studies are very difficult to elaborate until all influential parameters have been identified and defined. Even then, interfacial systems remain a formidable challenge to physical chemists. Just the separation of the influences of individual determining factors brought about by a single parameter change is often an elusive pursuit. Thus a change in the ratio of two immiscible solvents necessitates a change in the concentration of one reagent and alters the distribution of substances between the two phases. Alternatively, if the concentration is to remain unchanged, a change in reagent ratio is necessitated, which probably also changes the mixing efficiency, the solution viscosity, the precipitation point, the end groups, and other aspects of the polymer being formed. In some ways, though, there is a benefit to the experimenter in working with polymer synthesis, since molecular weight and polydispersity of molecular weight data add information to be reconciled with formulated mechanistic hypotheses. This advantage has certainly been proved in kinetic study of mechanisms and reactivity in free radical chemistry and in the fascinating study of Friedel-Crafts catalyst-cocatalyst systems. In Volume I of this book the reader will be introduced to some definitions of interfaces (Chap. 6) and given an experimental entree into mechanistic probing of interfacial polycondensations (Chaps. 5, 7, and 8).

Interfacial polycondensation may be classified as a low-temperature synthesis method. In contrast to the slow high-temperature polycondensation procedures, the low-temperature methods employ reactions that proceed at high reaction rates and are capable of giving quantitative yields at ordinary temperatures. Consequently, reagents of higher chemical potential must be provided than would normally be employed in a thermally driven reaction. The option also exists at low temperature whereby reagents of high chemical potential may be allowed to react when dissolved in a single phase (i.e., "solution polycondensation"), rather than by interfacial techniques. There are even further variations, such as the procedure in which reagents are separately dissolved in a common solvent and the solutions are initially layered. The reaction progresses, without stirring, at a boundary where the diffusing reagents meet. An additional special feature arises here if the product is insoluble and has integral tenacity; a new boundary and interfaces will develop at some time during the course of reagent conversion. These differences add significant variety to the preparation of a single polymer.

Applications of biphasic systems to the synthesis of small molecules are beginning to appear in the chemical literature (Chap. 4). Not included, but related to this theme, is the use of insoluble polymer supports in organic chemical synthesis, very recently reviewed by C. C. Leznoff. Expanded use of the techniques depends on disseminating information of successful applications.

One other subject, stirring, also has been included for the special benefit of young chemistry students. They are rarely taught that stirring may be a determining variable of the success or failure of a chemical reaction, the yield obtained, or the composition of byproducts.

Some advantages of good mixing, such as lower reaction time or lower reaction temperature, are illustrated in the following examples. Homogenizers have been used in the rapid aging of perfumes. The normal process, in which the perfume oils, alcohols, and the balance of the after-shave lotion formulation are mixed and aged to permit esterification reactions to take place among the components, requires 7 to 10 days, whereas this reaction is completed immediately by passage through a homogenizer. It is possible, by homogenizing a mixture

1. Introduction

of morpholine, water, and molten carnauba wax, to saponify some of the wax even though this reaction will not take place with ordinary agitation at the boiling point of water. The saponification of crude cottonseed oil in hexane and the preparation of soap greases in situ in the oil similarly profit from a decrease in the time and temperature requirement; the latter feature results in a furthered saving of time since the cooling cycle is reduced. In reported synthetic preparations one rarely sees mention of the quality of stirring in the way that reaction time, temperature, concentration, or pressure are given. Nor is an experimenter who elects to incorporate stirring usually aware of criteria that should govern his choices to produce effective mixing. Chapters 2 and 3 have been included to provide an introduction to a subject worthy of some attention.

Scale-up of an interfacial process sometimes presents unexpected difficulties. In the copolycondensation of aryl dicarbonyl and disulfonyl chlorides with bisphenols, Schlott and co-workers obtained water-in-oil emulsions that were not easily broken nor adequately freed of excess caustic and byproduct salts. As a result the polymers were of poor quality and stability. When the effect of stirring rate was examined it was discovered that the emulsified polymer continued to "grow" on standing, such that the initial low viscosities more than doubled in 30 min. Once established the emulsion is sufficiently stable to maintain the reaction unaided. The investigators turned to solution polycondensation for improved polymer quality. However, the large-scale work did provide the basis for a successful continuous polymerization reactor (Fig. 1.1), which rapidly produced large amounts of polymer emulsion.

An introduction to membranes and to projects involving bioorganic interfaces was included in this book (Chap. 10) to invite the chemist's attention to areas where he can contribute and to keep us all mindful of the fact that biochemistry is not an independent field, but one that has profited from and comes within the purview of organic, physical, and especially polymer chemists, physicists, and engineers.

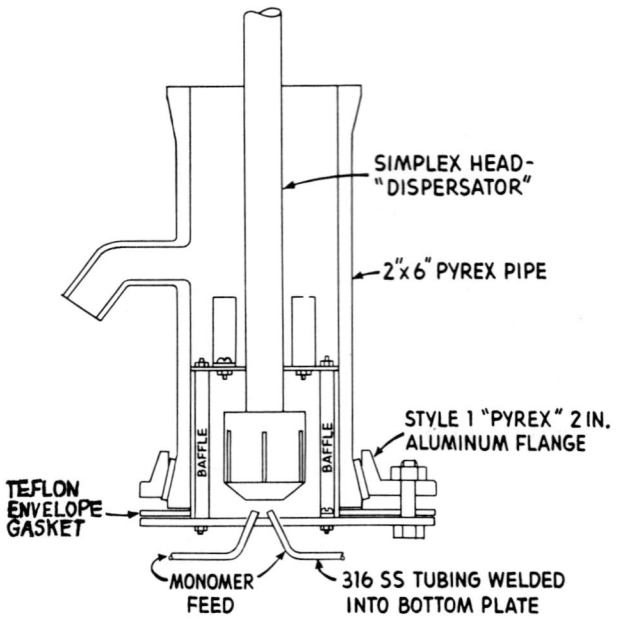

FIG. 1.1. Continuous reactor for interfacial polymerization. (R. J. Schlott, E. P. Goldberg, F. Scardiglia, and D. F. Hoeg, *Advan. Chem. Ser.*, 91:710 (1969).)

What economic factors operate for or against the use of interfacial polycondensation? This technique is thought of as a very convenient laboratory technique as it is usually conducted in a high-speed blender. Except, possibly, for reagent preparations the procedure is extremely rapid, usually involving water as an inexpensive ingredient with a relatively lessened fire hazard. High molecular weight polymers can be obtained without driving the condensation reaction to extremely high conversion and without the care needed in solution condensations for high purity and precise molar ratio balance of reactants. The polymer can often be isolated very conveniently, as by direct filtration of the reaction system. Because of product stability, some polymers can only be produced by low-temperature methods.

Generally speaking, a high-temperature polycondensation involves less expensive (more available) reactants than a low-temperature

1. Introduction

process wherein chemical potential must be built into a reagent, for instance, a carboxylic acid vis-a-vis a carbonyl chloride. High temperatures may produce contaminating byproducts, may induce reagent attack on reaction vessels (as happens with evolving acids or with the use of phosphorus compounds), and may be deleterious to the quality of the reaction product. On the other hand, high temperatures may produce a desirable low viscosity in a reaction medium, may accelerate reaction rates, and may volatilize undesirable byproducts or solvents used in heat exchange. Low-temperature interfacial polycondensations may suffer from problems with emulsification and contamination with occluded phase-transfer agents and reaction rate promoters. Low- and high-temperature polycondensations will produce products of different molecular weights, especially if precipitation of the product occurs during the reaction. These and other differences, the reader will find, distinguish the techniques in terms of macroscopic and economic characteristics.

On the face of it, interfacial synthesis would seem to be an economically favored procedural alternative. Yet industrial-scale production generally has favored monophasic reactions when a choice exists. Possibly the availability of fixed plant equipment, the familiarity with unit processes that have been established by years of experience, and the difficulty in scaling up high-speed stirring have been deterrents to the adoption of interfacial synthesis. Because data for analysis governing such a choice is difficult to find, the editors sought to have this subject examined (Chap. 11). It was also included because an economic analysis of a process is a valuable experience, which most chemistry students do not receive as part of their education until after they have left their halls of learning.

Areas in which interfacial synthesis is potentially important include some of the following applications. Greater use of the methods for the synthesis of "small" molecules may be made in the future as experience will distinguish the advantages of separating reagents, catalysts, and products among different phases and onto interfaces.

As is pointed out in several chapters, some materials cannot be synthesized in desirable yields except by a low-temperature method, or with a desirable, sufficiently narrow molecular weight dispersity. Industrial use of interfacial techniques for encapsulation may expand. Where industrial-scale solvent removal and recovery is of major consideration, especially in regard to energy consumption, interfacial synthesis—notably with gas-aqueous biphases—may hold economic advantages. Derivatization of commercially available natural and synthetic polymer molecules and surface treatment of fabricated products, adding desirable product properties, may well increase with experience. The preparation of dispersions directly applicable to product fabrication may be developed for condensation reactions in analogy with latex preparations from emulsion polymerization of vinyl monomers.

A workable mechanistic description of chemical reactions is always desirable. To have mechanisms extended to include system heterogeneity and real, practical complexities is an inviting and challenging realm of endeavor. It is hoped that an overview will be formed, forged by physical and chemical understanding, which would merge the islands of specialization and the waters between into a unified field of science.

2.

STIRRING IN ORGANIC CHEMICAL SYNTHESIS

*J. Henry Rushton**

*School of Chemical Engineering
Purdue University
Lafayette, Indiana*

I.	INTRODUCTION	10
II.	PRINCIPLES OF MIXING	11
III.	FLUID MOTION	12
IV.	EFFECT OF MIXING ON REACTION RATE AND EQUILIBRIUM	13
V.	MIXING	14
VI.	FLOW PATTERNS FOR MIXING	15
VII.	IMPELLERS	20
VIII.	FLOW FROM IMPELLERS	24
IX.	POWER	25
X.	TURBULENCE	27
XI.	CONSEQUENCES OF FLUID MECHANICS	28
XII.	GAS-LIQUID REACTIONS	31
XIII.	IMMISCIBLE LIQUIDS	32
XIV.	APPLICATION OF PRINCIPLES	36
	REFERENCES	36

---

*Also technical advisor, Mixing Equipment Co., Rochester, New York.

I. INTRODUCTION

It is the purpose of this chapter to define the effects of stirring and to show how to control them for optimum benefit in organic chemical reactions. The underlying mechanisms are found in the science of fluid mechanics. Fluid motion encountered in reaction research and industrial-scale work will be discussed.

Many organic reactions occur at rates which are such that fluid motion can play a major role in both reaction rate and yield of desired product. It is well known that different rates of stirring can significantly affect product distribution and the control of reaction rate. It is, therefore, important to recognize what a stirring device does to a fluid and how this, in turn, affects the chemical reaction. Stirring moves fluid, and it is this motion together with the chemical kinetics of the reaction which interact to produce the combined effect.

In engineering practice stirring is most frequently referred to as mixing. Mixing of fluids is ordinarily done to intermingle parts of a mass to achieve, or approach, a homogeneous distribution of substances. Likewise, mixing is used to achieve a distribution and control of heat content and temperature. Both of these objectives may have a pronounced effect on reaction kinetics. When immiscible phases are present, either liquid-liquid or gas-liquid combinations, the fluid motion during stirring may also be used to generate drops or bubbles which result in interfacial surfaces where material is transferred and reactions may take place. Even in these instances, however, the fluid motion will result in mixing the continuous phase in addition to developing interfaces. Thus, mixing motion can enhance the delivery of and removal of material to and from the reaction site; chemical diffusion rates may be altered by the "eddy diffusion" or turbulence in the moving fluid. Stirring, or mixing, can be thought of as an operation whereby the reaction surface can be enlarged and controlled, and forced convection can be applied through turbulence to achieve more rapid transport of molecules than can be accomplished by diffusion alone.

2. Stirring in Organic Chemical Synthesis

Although the research chemist may not be thinking in terms of large-scale industrial application, it is well for him to remember that the science of mixing has been developed because the problems in scaling up laboratory results have given the incentive necessary to understand and to use the principles of fluid mechanics to reproduce the research laboratory results on a large scale. As a result, it is to the advantage of the research chemist to understand the fluid motion of stirring and how these kinetics will interact with the chemical kinetics. It is not sufficient for a research chemist to produce a result in one particular type of apparatus. Rather, he should understand the fluid kinetics involved so that he can define a fluid motion to give the best result, so that his result can be achieved both on a smaller or larger scale. Such insight involves a more complete understanding of the overall kinetics of the synthesis.

II. PRINCIPLES OF MIXING

Mixing of liquids can be considered in two parts: (1) the hydraulic regime and energy required for the desired results for the particular process, and (2) the flow regime and energy production and distribution supplied by the mixing device, i.e., the impeller in stirring operations. The first part is a process requirement which is studied in the research laboratory and pilot plant. The literature offers very few specific data of process requirements for mixing in organic reactions; these are determined in the laboratory for each reaction. Apparatus for research of this kind should be set up in such a way that the best flow pattern and optimum mixing-energy input can be determined for a particular system.

The second part is a knowledge of performance of the mixing device so that the process requirement can be met. Impeller performance is dependent upon many variables: impeller shape, size, position, and speed of rotation; container size and shape, depth of liquid, and shape and size of internal fittings such as baffles and coils; fluid density, viscosity, flow rates, and other physical properties. The ratio of impeller size (diameter) to container size (diameter) may vary from 1/4 to 1/2, and the most common ratio is about 1/3. For side-entering propellers in large tanks the ratio may be as low as 1/100.

Considerable data are available on impeller performance for large-scale operations [1-6], and mixing techniques in the organic laboratory should be of the same proven type used in such operations. A wide variety of homemade impellers and mixing devices are often used. While the devices operate effectively, further experimentation is usually required to reproduce the results in large batches or in other containers. The laboratory worker will be able to eliminate mixing as an unknown variable and use it with confidence as a tool to control a reaction if he pays more attention to the known facts concerning the behavior of impellers and other mixing devices, the properties and dynamics of fluids, and the shape of containers. He would also be assured of being able to reproduce his work on a larger scale, as far as the mixing problem is concerned.

III. FLUID MOTION

A mixing impeller is used to move fluid; it of itself does not mix material, it serves only to move the fluid at different velocities and directions from the surrounding fluid. Mixing is accomplished by entraining fluid into the expanding stream flowing from the impeller. The mechanical factors that determine the fluid regime in a mixing vessel are the container shape, the impeller and its position, and the presence of any stationary object in the liquid. To produce interchange of particles in a mixing apparatus in the shortest time, it is essential that currents of flow cause both vertical and horizontal motion. Sometimes intense vertical currents are desired, for example, when heavy solids are to be suspended in liquids. At other times lateral or horizontal flows are required, as is the case when gases are dispersed through liquids, or when miscible liquids of approximately the same density are to be blended. To obtain the effect best suited to a particular operation, it is necessary to know how each of the mechanical factors (container,

2. Stirring in Organic Chemical Synthesis

impeller, and stationary objects) exerts its influence in the fluid-motion pattern.

For effective mixing it is usually necessary to avoid vortices at the liquid surface to obtain the type of flow that ensures vertical and lateral mixing. Whenever the combination of container, impeller, and stationary objects results in deep vortices, liquid swirl predominates in various degrees and results in the minimum of vertical and lateral currents of flow. Swirling liquid produces laminar circular flow. Particles of like-densities traveling in such streams are moved away from the center of the motion by centrifugal force and will stay away from the central portion of the mass. This may result in separation, or classification and thickening, rather than in suspension and mixing.

Mass flow of fluid and turbulence both combine to produce forced convection and mixing. Large-scale mass flow results in a major current and pattern of flow, resulting in the transport of material over long distances. Small-scale eddy currents, such as are present in turbulent flow motion, move material over short distances, in all directions and provide small-scale mixing. This is the "agitation" mechanism of mixing. The turbulent eddies are carried throughout the fluid by the mass flow.

The proper combination of mass flow and turbulence is necessary for the most economical use of energy for mixing. The amount of flow and turbulence are functions of the impeller speed and size, of the container size and shape, and of the physical properties of the materials to be mixed. The mixer produces only mechanical fluid-flow effects. Any chemical effects are the results of mass transfer to interfaces where reactions occur.

IV. EFFECT OF MIXING ON REACTION RATE AND EQUILIBRIUM

The highest reaction rate that can be reached without mixing is a function of the molecular species involved, of their concentrations, and of the temperature. In heterogeneous reactions the actual rate will be inversely proportional to the length of path of travel of the reactants to the interface. It will drop as the viscosity of the medium increases. While a mixer may be used to increase the area for reaction and to decrease the path for diffusion, it is not yet possible to predict the optimum impeller discharge and turbulence to obtain the desired rate of a reaction. Hence, mixing variables must be evaluated experimentally by the research chemist.

V. MIXING

Heterogeneous reactions can be classified into three groups, in which

1. The chemical reaction rate is greater than the rate of diffusion to the zone of reaction, with the result that the observed rate is controlled by the diffusion rate
2. The chemical reaction rate is less than the rate of diffusion, in which case the observed rate is the chemical reaction rate and is controlling
3. The chemical reaction rate and the diffusion rate are of the same order of magnitude, in which case the net rate is dependent on both

In each of the three groups the area of contact between the heterogeneous portions may be affected by agitation. For example, a granular solid resting on the bottom of a container may not present maximum surface to a solvent because of surfaces of solid which touch each other. If the solid is suspended by means of agitation, a greater area of solid-liquid contact will be obtained and the overall rate of reaction will be increased. When gases are introduced into liquids, a mixer can be used to produce small bubbles and thus to extend the area of contact for reaction; in this case also the mixer will influence the rate of reaction regardless of which of the three classifications applies.

If the reaction belongs to class 1 or 3, mixing may affect the rate by supplementing the diffusion process. When fluid flow is produced by a stirring device the position of reactants is altered and a greater degree of homogeneity is obtained. The forced convection, both in large masses and in small eddy currents, supplements the normal diffusion of reactants and can also be used to offset gravitational effects.

With most heterogeneous systems mechanical mixing is essential in order to reach equilibrium conditions in a short period of time. The difficulty in obtaining equilibrium increases with the size of the batches, and with larger batches it is often difficult to reproduce the true equilibrium conditions since the conditions may have been determined on a much smaller scale. Side reactions occur often in organic processes, and close regulation of temperature and concentration is necessary to control their extent. If the optimum conditions for a certain reaction have been established, uniformity of these conditions throughout the mixture helps to limit side reactions to a predictable minimum. In many cases a mixer may be employed to achieve the desired result.

Frequently, as a result of mixing of the system, the reaction rate increases so much that the heat of reaction is liberated faster than it is removed through cooling surfaces or evaporation. A temperature increase results which, in turn, speeds up the reaction. The rise in temperature may also shift an equilibrium to a less desirable condition and may stimulate undesirable side reactions. On the other hand, mixing will assist in the convection of heat and thereby under proper conditions afford a better chance to maintain a desired uniform temperature through better heat transfer.

VI. FLOW PATTERNS FOR MIXING

Organic reactions are ordinarily carried out in the laboratory in glass flasks of various shapes or in glass or metal beakers. Stirring is done with small flat paddles or with various shapes of propellers. The flow motion ordinarily desired is one that will eliminate swirling motion, and the desired flow patterns should have vertical and lateral flow motion. The pattern of fluid motion depends upon the impeller, the fluid, and the container. Since all three interact to produce the final result, the interaction is rarely a simple relation.

Ordinarily, impellers are classified as axial, radial, or mixed-flow types. The marine type is an example of an axial-flow impeller; a flat paddle and a flat-blade turbine are illustrations of a radial-flow impeller; and a pitched-blade or axial-flow turbine illustrates a mixed-flow or combination axial- and radial-flow impeller. These descriptive terms, however, are accurate only when the tank (and its fittings) and the fluid together allow the impeller to produce the desired flow [1-4]. For example, for any single rotating impeller—regardless of shape—located in a vertical cylindrical tank, with a vertical shaft, there will be a preponderance of rotary swirling or tangential motion. Here the propeller and the turbine will give substantially the same flow pattern. If, however, in the same tank the low-viscosity liquid is replaced by a very viscous one, the propeller will give essentially axial flow and the turbine will give essentially radial flow. If for the low-viscosity liquid the configuration of the tank is changed by the addition of baffles at the side walls, the propeller becomes an axial device and the turbine a radial device. In the case of side-entering propeller mixers, where the shaft is horizontal, swirling will always result unless a particular off-center position is chosen and will also be dependent upon the direction of the propeller rotation. It is possible to destroy swirling and rotary motion by proper off-center placement of top-entering shafts that are at an angle to the vertical. All these flow patterns are illustrated in Figs. 2.1 to 2.5.

Organic laboratory experiments are usually carried out in beakers, bottles, and round-bottom or Erlenmeyer glass flasks. Such containers are almost the poorest conceivable for obtaining good mixing conditions because a rotating impeller will cause a swirling flow which does not permit rapid mixing. Even drugstore milk-shake equipment is more suitable for mixing than most laboratory glassware, for the container has indented sides which provide baffling and thus eliminate some of the swirl. An excellent type of glass container for organic laboratory use is the creased flask shown in Fig. 2.6. Creases are pressed inward in the sides of ordinary or three-necked

2. Stirring in Organic Chemical Synthesis

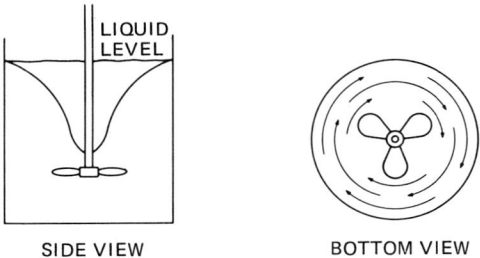

FIG. 2.1. Swirling flow pattern for impeller of any shape, without baffles.

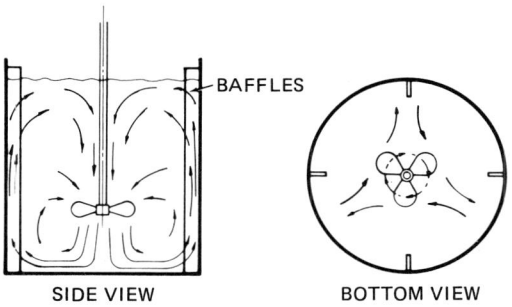

FIG. 2.2. Flow pattern for propeller with baffles at tank wall.

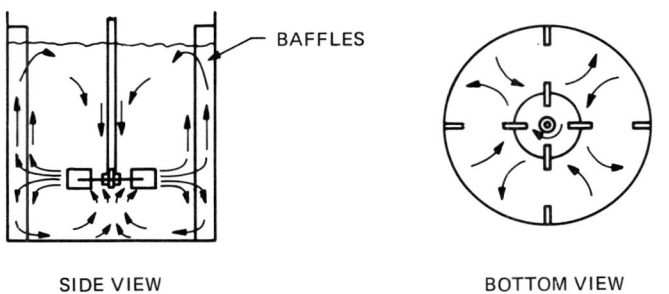

FIG. 2.3. Flow pattern for turbine with baffles at tank wall.

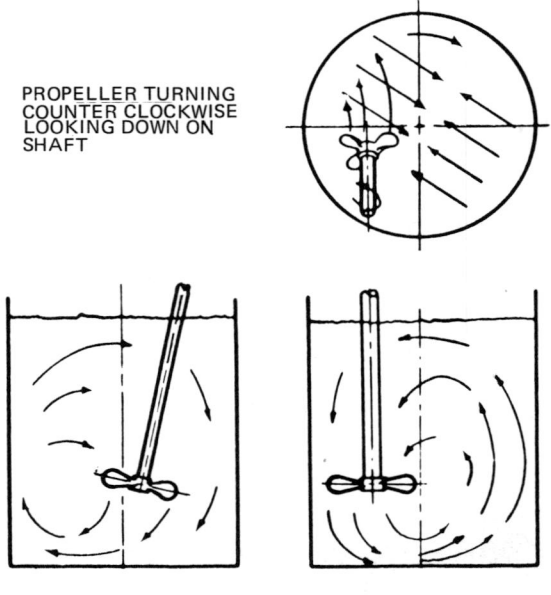

FIG. 2.4. Flow pattern for propeller top-entering, off-center, without baffles.

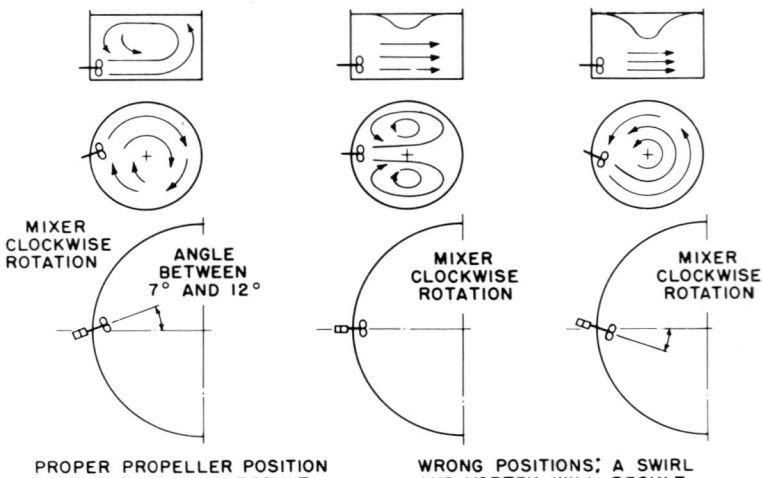

FIG. 2.5. Side-entering propeller mixer position.

2. Stirring in Organic Chemical Synthesis

FIG. 2.6. Propeller operating in flask with creases.

flasks. A typical 2-liter round-bottom flask has four vertical, equally spaced creases, each 7.5 to 10 cm long, 0.6 to 2.5 cm deep, tapering in width from 0.3 to 3.75 cm at the circumference. The creases act as vertical baffles, break much of the swirl, and produce lateral and vertical flow currents, which can be duplicated in larger vessels. These flasks provide excellent mixing conditions under which the ratio of flow to turbulence can be controlled; they are standard laboratory articles.

In the absence of creases, vertical obstructions may be mounted inside a flask or bottle to act as baffles, or the impeller may be placed off-center with the shaft at an angle to the vertical.

VII. IMPELLERS

Rotating impellers are most often used for mixing purposes, and the present discussion is therefore restricted to propellers, rotating paddles, and turbines. Reciprocating and vibrating devices are, however, employed in some cases.

Innumerable impeller shapes can be used to move fluid, but long experience has shown that a few shapes can reproduce all the results obtained from a wide variety of shapes. By focusing attention upon these few shapes one can study mixing from the standpoint of fluid motion, realizing that the entire function of the mixing impeller is simply to move fluids. Shear stresses are developed when a high-velocity discharge stream rubs against adjacent fluid. Shear between an impeller and the fluid is negligible compared with fluid shear; fluid touching the impellers is moving at the volocity of the impeller, hence there is a minimum of shear at the impeller surface.

The three impeller shapes that have received the greatest amount of attention and study are the three-blade marine type of propeller, the flat- or curved-blade turbine, and the flat paddle. Propeller blades are so shaped with respect to angle and corresponding radius that they have constant pitch. Turbines are commonly made by mounting blades of various shapes and sizes on a circular disk. (Figure 2.9 shows a standard flat-blade turbine.) The flat paddle, which was the first and most common rotating mixing device, is used less frequently in modern industrial practice. However, because of its simplicity it is the most commonly used impeller in organic synthesis laboratory work. It is ordinarily either a single flat blade mounted on a shaft attached to the center point of the blade or two blades mounted in the same way so that there are four paddle arms. The flat face of the blade may be either parallel to the shaft axis or at an angle to it: in the latter case it is spoken of as a pitched-blade axial-flow turbine.

2. Stirring in Organic Chemical Synthesis

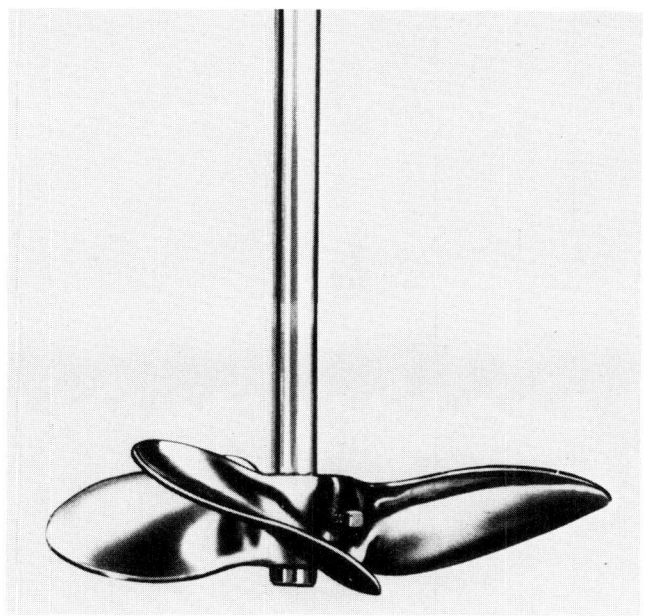

FIG. 2.7. Three-blade marine type, axial-flow propeller. (Courtesy Mixing Equipment Co.)

The simple three-blade marine-type propeller shown in Fig. 2.7 is a very common and useful impeller for laboratory or large-scale operations. In a properly baffled tank, such a propeller produces axial flow, i.e., fluid will be displaced downward along the axis of rotation, and a top-to-bottom-to-wall-to-top flow pattern will result, as illustrated in Fig. 2.2. The core of liquid below the impeller will rotate, but resistance of liquid, walls, and baffles converts most of the motion to horizontal and vertical flow as shown in the figure.

Paddles (Fig. 2.8) and turbines (Fig. 2.9) with the proper baffles will produce radial flow, i.e., flow directed outward from the blades in a plane at right angles to the axis of rotation. The general flow pattern produced by such impellers is illustrated in Fig. 2.3. The predominate flow curves out from the blades to the

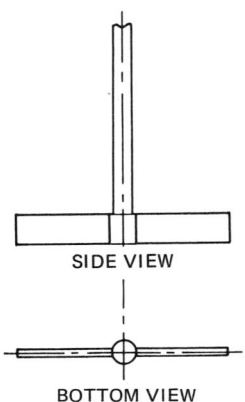

FIG. 2.8. Two-blade flat paddle.

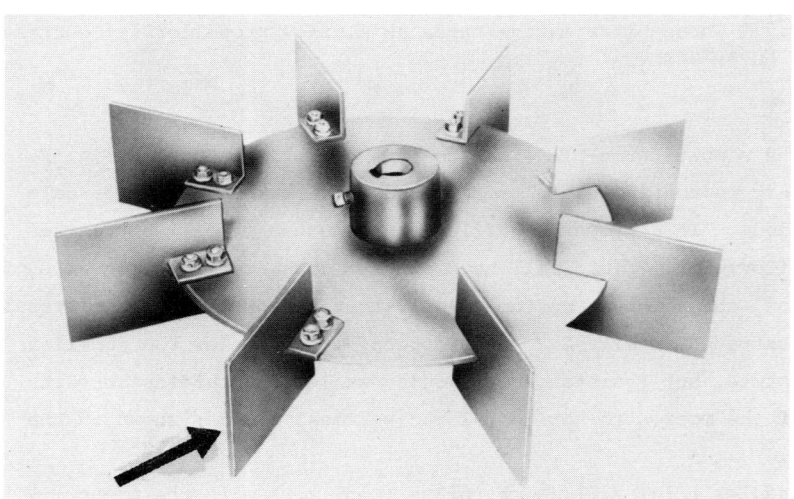

FIG. 2.9. Flat-blade turbine. (Courtesy Mixing Equipment Co.)

2. Stirring in Organic Chemical Synthesis

side of the container and then splits, part moving down and part moving up, and then both parts turn toward the impeller, thus completing the pattern.

There are many variations of these principal impeller types. While flow from any modification of the marine propeller is usually predominantly axial and that from paddles and turbines is radial, mixed flow made up of both types is obtained in some cases. However, when any impeller, whether a propeller, a paddle, or a turbine, rotates in a smooth, vertical, cylindrical, or spherical container without stationary objects and in a low-viscosity medium, the rotary motion of the impeller will carry throughout the fluid and no appreciable axial or radial flow will take place. The fluid will simply swirl, a deep vortex will form, as shown in Fig. 2.1, and the minimum of agitation and mixing will result from the energy supplied. There is one important exception to this, namely, an axial-flow propeller in off-center position. By using creased flasks or cylindrical containers with appropriate baffling, it is possible to achieve a good top-to-bottom turnover and flow pattern, and also to supply energy in the most effective way.

The ordinary glass paddle can provide adequate stirring for many laboratory operations, especially in creased or baffled flasks which transform the radial flow into the vertical currents so essential for adequate mixing. A small metal propeller is an excellent laboratory device. Used in the proper container it produces vertical-flow currents, and swirl can be eliminated by the use of baffles or off-centering. Ordinarily, the impellers should be located about one-third flask diameter or about one impeller diameter above the bottom of the flask. If the liquid depth is great, two impellers placed at least one impeller diameter apart can be used on the same shaft. Propellers or any other axial-flow impellers should be used to discharge downward so that the fluid stream is deflected by the bottom of the container toward the walls. This is true whether the impellers are used singly or in pairs.

If emulsions are to be formed, or if solid particles are to be broken or shredded, high-speed stirring is often required. In the laboratory, small-diameter metal propellers of the type previously mentioned can be used up to speeds of 10,000 rpm, drawing up to 0.08 hp. Such speeds, however, are rarely used in pilot-plant or larger-scale operations. To attain high shear in large units, large-diameter impellers are used to attain an impeller-tip speed comparable to that occurring in the small-diameter, high-speed laboratory mixers.

As mentioned above, swirl may be avoided without the use of special containers or baffles by placing the propeller in an off-centered position, as illustrated in Fig. 2.4. The shaft enters the surface of the liquid to one side of the center line and passes down through the liquid at an angle to the vertical. The exact position and angle of the shaft to secure agitation without forming a vortex depend upon the size of the vessel and the depth of the liquid. The position is critical, and a variation of shaft angle, propeller depth, or lateral placement may allow swirling to take place. Top-entering off-centered axial-flow propeller mixers are available in small sizes for laboratory use and are employed in manufacturing operations in sizes up to 25 hp. They are usually applied to low-viscosity liquids; the propeller is immersed at least one propeller diameter below the liquid level and operates at relatively high speed. Use of more than one impeller on a shaft is necessary when the liquid depth is greater than the container diameter.

VIII. FLOW FROM IMPELLERS

To characterize the effect of stirring, the mixing impeller characteristics should be known for the desired flow pattern. The important characteristics are: the quantity of flow produced, the velocity of flow, the turbulence in the flow, and the power transmitted from the impeller to the flow. It is the interrelation among these quantities which results in various fluid flow regimes that will best fit the requirements of the chemical kinetics.

2. Stirring in Organic Chemical Synthesis

The quantity of liquid moved by an impeller is a function of impeller shape and size (diameter) and of the speed of rotation, as well as of the shape of the container and its fittings. The flow is defined as the amount of fluid which moves axially, or radially, away from the impeller at the surface or periphery of rotation. For example, the flow from a marine-type propeller is the amount of fluid per unit time which moves in an axial direction from the circular cross section described by the rotating blades. For the flat-blade turbine the flow is that which moves radially through the ring-shaped area described by the periphery of rotation of the impeller.

The relation between flow, speed, and size is

$$Q = K_1 N D^3 \qquad (2.1)$$

where Q is the volumetric flow per unit of time, K_1 is a constant for the impeller type and its proximity to walls and liquid surface, N is the impeller speed in revolutions per unit time, and D is the impeller diameter. Values of K_1 vary from 0.4 for a square-pitch propeller, to 0.5 for flat paddles (two blades), to 0.7 for a six-blade standard flat-blade turbine [5].

IX. POWER

It is important to remember that stirring requires work, and this is referred to per unit of time as power. Thus, to compare the action of stirring or mixing devices, the comparison should be made on the basis of equal power applied through the impeller.

The power required to turn an impeller in a fluid is exerted by the impeller on the fluid. Ultimately the power is completely absorbed through friction in viscous and turbulent shear stresses and is dissipated as heat.

The amount of power (P) required to turn an impeller depends upon its shape, size, and speed, the density and viscosity of the fluid, and the shape and fittings of the container. The position of the impeller, the depth of the liquid, and the presence or absence of a free liquid surface also affect power. The basic relations are:

$$P = (K_2/g_c)\rho N^3 D^5 \quad \text{(fully developed turbulence)} \tag{2.2}$$

and

$$P = (K_3/g_c)\mu N^2 D^3 \quad \text{(viscous conditions)} \tag{2.3}$$

where g_c is the gravitational constant, ρ and μ are the density and dynamic viscosity of the fluid, and N and D are the speed and diameter, respectively, of the impeller as defined above.

The proportionality constants K_2 and K_3 are for turbulent and viscous conditions based on Reynolds numbers. Their values are of minor importance in this discussion, and the reader is referred to other works on mixing [4-6]. For orientation a few typical values of K are as follows: for a standard six-flat-blade turbine K_2 is 5.5 and K_3 is 62.0; for a square-pitch propeller K_2 is 0.32 and K_3 is 41.0; for a flat paddle (blade length D of 4 times blade width) K_2 is 2.25 and K_3 is 43.0.

Power data have been correlated on the basis of the dimensionless Reynolds number and power number (equivalent to the drag coefficient). The Reynolds number used for mixing is $D^2 N \rho / \mu$. The power number is $P g_c / \rho N^3 D^5$.

At high Reynolds numbers, where there is no swirling motion, it is postulated that turbulence is fully developed and that viscous forces do not control power requirements. At low Reynolds numbers it is postulated that viscous forces are controlling and that the flows are streamlined and viscous.

Power is equal to flow (Q) multiplied by pressure (or head, H) against which the flow moves. This is a useful concept in mixing because the power applied to a fluid by a mixing impeller produces both flow and turbulence. Turbulence is directly proportional to the head developed by the impeller [4]. Thus, power (gram-centimeters per second) can be thought of as equal to flow (grams per second) times the total potential head (centimeters) against which flow takes place:

$$P = Q\rho H \tag{2.4}$$

2. Stirring in Organic Chemical Synthesis

It is exceedingly difficult to measure directly the head produced by open-running mixing impellers (as distinct from a confined impeller such as in a centrifugal pump); however, it is easy to calculate the head from Eq. (2.4).

X. TURBULENCE

Since turbulence in a flowing stream from a mixing impeller is directly proportional to the head developed by the impeller, head is used as an index of the level of turbulence.

Turbulence consists of eddies which are characterized by size and velocity. Eddies, however, have the property of continual change. Some eddies grow in size and decrease in point velocity, whereas others decrease in size and increase in point velocity. Finally, all eddies transfer mechanical energy into heat and are thus dissipated. Eddy size is a function of the system size, and larger systems permit larger eddies; eddies vary from largest size to zero. Present knowledge allows definition of turbulence only as the product of eddy size and velocity. This is spoken of as eddy viscosity, or turbulence. Hence, even for the same mathematical number used to describe turbulence in two different-sized systems, there may well be a difference in the velocities of turbulence (for the same fluid velocity) at corresponding points in the two systems, compared with the scale or size of the turbulence in these two systems. The only existing quantitative information regarding turbulence in a mixing tank is that computed by dividing power by flow. This, however, is very useful even without complete knowledge of individual eddy viscosity and turbulence.

To appreciate the significance of the relations just shown, note that power is proportional to flow times head; and since flow, Q, is proportional to ND^3 (Eq. 2.1), and power, P, is proportional to N^3D^5 for turbulent conditions, it follows that

$$H \propto N^2 D^2 \qquad (2.5)$$

For viscous flow conditions, by the same reasoning and using Eq. (2.3),

$$H \propto N \tag{2.6}$$

For the same power to be exerted on a fluid, it is clear from Eqs. (2.2) and 2.3) that a smaller diameter impeller must run at a higher speed. Thus, the flow will become less and the turbulence (or head) will become larger for this smaller impeller (of geometric similarity). And, conversely, larger geometrically similar impellers will produce larger flow to turbulence ratios at constant power than smaller impellers. This is one of the most significant aspects for determining the effect of stirring on an organic reaction.

XI. CONSEQUENCES OF FLUID MECHANICS

Consider a reaction in a baffled (or nonswirling) mixing situation where the diameter (D) of a paddle impeller is one-third the diameter (T) of the container, or $D = T/3$. If a smaller impeller is used and operated at a speed N to exert the same power in accord with Eq. (2.2), and if the same temperature is maintained, a difference in reaction rate or yield or other effect may (and frequently will) result. This difference in result can then only be attributable to the fact that the flow to turbulence ratio has been lower than where the larger impeller was used. Again, if an impeller larger than $D = T/3$ is used, the result may again be different, and for the same reason. In this way the type of flow regime best suited to the reaction can be determined. If larger impellers in the same-size container and same liquid amount give better results at the same power, it shows that the quantity of flow is controlling and that turbulence is of less importance than flow. Conversely, if better results are achieved with smaller impellers, it shows that turbulence and shear stress in the liquid are controlling and that flow is of less importance.

Impellers of the same diameter but of different shape will give different ratios of flow to turbulence when operated at the same power. The spectrum of flow and turbulence at constant power for a

2. Stirring in Organic Chemical Synthesis

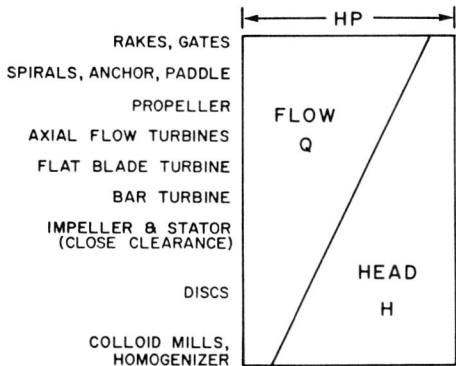

FIG. 2.10. Impeller spectrum.

wide variety of impellers is shown in Fig. 2.10. Here the relative flow (Q) and turbulence (head, H) for various impellers is illustrated graphically. For example, gates and rakes give the largest amount of flow and the smallest amount of turbulence. A flat paddle also gives high Q/H ratios. The flat-blade turbine is at about the midpoint of the flow-to-head ratio. Colloid mills and close-clearance homogenizers give the maximum turbulence, and thus the lowest Q/H ratios. This impeller spectrum is qualitative, and it is possible to modify the various impeller types (for example, by changing the angle of the blades or their proximity to walls and other surfaces) to change the Q/H ratios slightly. However, it can be seen that impeller shape is sometimes important to give the Q/H ratios desired.

It should be noted here that when high turbulence is required to produce a large interfacial area, it may not be possible for large impellers to give the sufficiently low ratios of Q/H required. In such cases colloid mills are used either alone or, frequently, in combination with propeller or turbine mixers. For example, some saponifications are best made by accurate feed of the several streams directly into a colloid mill, rather than into a mixer using a propeller in which long holding times are required for the reaction. To homogenize and stabilize wax and water systems, mixes can be made in a holding tank with turbines or propellers, and feed from the tank can be taken to a colloid mill to achieve the small particle size necessary. Thus, various combinations can be used in large equipment to provide conditions that cannot be achieved by a single unit with the same energy economy.

There are some reactions that will not be greatly affected by different ratios of flow to head. Also, it is sometimes desired to operate at a high power input so that the effect of changes in mixing rate will not affect the overall kinetic rate. This is accomplished by turning the impeller at speeds above which no effect is evident due to increased speed.

Note that power is proportional to the cube of impeller speed, no matter what the shape of the impeller. Further, all the power required to rotate the impeller is dissipated finally as heat to the liquid. Excessive use of power may well cause heat transfer problems and temperatures may reach too high a level.

The energy put into the system by the stirring device is used in the forms of flow and turbulence, produced by the shear stress of the flow from the impeller moving through surrounding fluid. It is the combination of these two which interacts with molecular diffusion to move molecular species to and from a reaction site. Attention to these fluid motion parameters will allow the organic chemist to provide the best stirring conditions for his particular reaction.

Certain types of reactions are known to be flow dependent, and others are turbulence dependent. For example, blending of fluids, homogeneous reactions, heat transfer, suspension of solids, and dissolving are examples favored by high ratios of flow to turbulence. Reactions involving immiscible liquids, and gas-liquid reactions, are usually favored by high ratios of turbulence to flow. In fact the interfacial area of drops of one phase dispersed in another will first increase rapidly as the ratio of impeller diameter (D) to container diameter (T) is increased. But, at a ratio of D/T of about 0.25, the maximum interfacial area will be reached. Further increase in the D/T ratio will result in a decrease in the interfacial area.

This holds for constant power input. Higher levels of power can of course be used to achieve a large interfacial area when larger impellers are used (see Fig. 3.1).

XII. GAS-LIQUID REACTIONS

When a gas is to be dispersed in a liquid it can be accomplished by feeding the gas to the "eye" of the impeller, or through a sparge ring below the impeller, or by entraining the gas through the surface of the liquid.

The type of impeller best suited for rapid dispersal of gas to achieve small bubbles and large interfacial area is the flat-blade turbine. The gas is fed from below the turbine in an upward flow at the centerline of rotation. Alternatively, the gas may be dispersed by a ring below the edge of the turbine, with the gas emerging through holes which direct the flow upward or horizontally into the stream flow.

Liquid-gas operations are complicated by the fact that there are often three separate physical steps in the process to be accomplished. First, a component in the gas must diffuse and pass through a gas film to the interface. Second, the component must pass from the interface through the liquid film and into the main body of liquid. Third, the components must react with another material in the liquid. The first two steps are governed by mass transfer rates and depend on interfacial area, concentration driving forces, and turbulence in the gas and liquid, respectively. The third step is chemical reaction, or sometimes absorption.

The process can of course be reversed, as when a chemical reaction produces a material that then passes from the liquid to the gas. The mixer, which produces the major effect on fluid motion, is then effective in providing some turbulence in the gas bubbles and a large amount of turbulence in the liquid, as well as causing dispersion and large interfacial area. The rate controlling step will be the slowest one of the three.

If the gas film mass-transfer step controls, then the application will be designed around the gas phase requirements, and the other two will be sufficiently provided for. When the liquid film mass-transfer step controls, the mixer specifications will be designed on that basis. Since the impeller works directly on and through the liquid, it is much easier to impose varying amounts of turbulence on the liquid, and thus affect the liquid film transfer coefficient, than it is to affect the turbulence in the gas bubble. For both of these mass-transfer controlled steps, high interfacial area is desired.

The chemical reaction becomes the controlling rate whenever the mixer has produced sufficient motion to increase the gas film and liquid film transfer rates to higher levels than the specific chemical rate. When this condition is reached further mixing power is unnecessary. As a practical matter, the power necessary to provide adequate interfacial area and good gas-bubble distribution is almost always enough to achieve these high overall film transfer rates.

VIII. IMMISCIBLE LIQUIDS

When two immiscible liquids meet at the mixer impeller, shear stresses occur in the outlet stream which produce drops of one phase within the other. Equilibrium is established between breakup and coalescence as the mixture circulates in the tank. The result is to produce a range of drop sizes.

Average drop size and size distribution will depend upon the mixing variables and the flow pattern as well as the physical properties of the two liquids.

Many types of immiscible liquid systems are encountered in organic syntheses. All these systems are mixed for specific purposes, for reaction, for solvent extraction, for color removal or clarification, for the removal or addition of heat, and for modification of mass transfer rates in reactors. The more common systems are water and hydrocarbons, polymerizations, and acidic or alkaline solutions to be combined in organic liquids.

2. Stirring in Organic Chemical Synthesis

The terms emulsion and dispersion are often used interchangeably. However, a dispersion is a general term implying distribution, whereas an emulsion is a special case of dispersion with a stable nonsettling and non coalescing distribution of colloidal-sized drops of one liquid in another. A familiar example of an emulsion is homogenized milk— an oil in water emulsion.

Production of the small drop size required for an emulsion generally calls for much higher impeller speeds and power input than for a dispersion. However, chemical reagents are frequently added to give surface effects, to lower the interfacial tension, and to allow production of smaller drops. Very little stirring is required to produce large interfacial areas in such cases.

A dispersion is a dynamic mixture that exists because energy is continuously supplied to the mixture itself. Since the droplet size will be larger than that prevailing in an emulsion, considerable less power is required for the desired result.

The dispersed phase may be either of the two fluids, regardless of the volume ratio of them. Generally, if a radial-flow impeller is located just below the interface of the two liquids at the start of the mixing process, the lighter or upper phase will be drawn into the lower phase and dispersed in it. This dispersion will exist only as long as the mixer is in operation at the proper speed. A speed change can often result in a phase inversion.

There is an economic limit to the minimum drop diameter obtainable with a particular mixing setup. Correlations between drop size and power input indicate that a small drop size or a large interfacial area can be achieved only at the expense of relatively high power consumption.

With approximately uniform mixing turbulence throughout the tank, the suspended droplets will be subdivided until they reach an equilibrium drop size as a result of two opposing mechanisms: drop coalescence and drop breakup. The intensity of turbulence will not be uniform throughout a tank. In those areas where less turbulence occurs, colliding droplets will coalesce to form larger droplets.

When a larger droplet is transported to an area of higher turbulence, it will be broken up.

A properly designed fluid reaction vessel for an immiscible liquid system will establish a dynamic equilibrium to control droplet size and interfacial area to the greatest practical degree.

The design of a correct mixer for any immiscible liquid system depends on proper selection of the impeller and on the fluids themselves, the vessel shape, and the end results to be obtained. To determine the "correct" mixer for an immiscible system, the factors given below should be considered.

An impeller producing a high shear rate will provide low flow compared to an axial-flow impeller at the same power. Again, note the impeller spectrum chart given in Fig. 2.10. For equal horsepower, a high shear-rate impeller will give considerably less flow than a high-flow helical impeller or anchor impeller.

Since many reactions depend on effective mass transfer across liquid-liquid interfaces, it is important to obtain the maximum practical interfacial area. The selection of the impeller and speed will vitally affect the drop size and dispersion and, accordingly, the interfacial area. However, if the drop size is too small the dispersion may be difficult to separate.

Various immiscible liquid systems will exhibit different responses to flow and to fluid shear stress or turbulence. Obviously, this means it is important to determine which is critical (flow or turbulence) and the acceptable ratio of flow to turbulence (or head).

The ultimate stability of an emulsion is determined by the interfacial tension, drop size, and chemical ingredients and is favored by low interfacial tension. Emulsification is ordinarily obtained in special equipment that is designed to produce high liquid shear stresses.

Homogenizers, which produce high shear stress and high turbulence, are used for rapid emulsion formation. For example, in a saponification reaction when an alkali solution is to be added to an oil, the addition is best made as near the eye of the impeller as

2. Stirring in Organic Chemical Synthesis

possible. This will assure the most rapid generation of small drops and distribute them quickly in the impeller flow. This is true whether a common type of mixing impeller is used or a homogenizer or Waring blender-type unit is used. Whenever large surface areas are desired in gas-liquid or liquid-liquid interfacial systems, the added phase should be added as near the eye, or center of the impeller, as feasible. Rapid distribution and uniformity of area will promote rapid reaction and result in most efficient use of the chemicals.

If a mixer produces a dispersion where some drops are smaller than the stable drop size, these drops will coalesce to a higher average size. If the mixer is producing stable drop sizes, the emulsion will remain in this stable condition.

Impeller location is vital in any dispersion action in immiscible liquid mixing. Depending upon the fluid properties of the two phases, it may be possible that only one phase can be dispersed in the other, or it may be possible to disperse either phase. If only one phase can be dispersed in the other, then the impeller can be located in either phase. If it is possible to disperse either phase, then starting a batch operation with the impeller in the desired continuous phase will make that phase the continuous one.

If, in a batch operation, dual impellers are used at low mixer speeds and power levels, and one is used in each phase, two separate dispersions would be obtained. At high horsepower levels, however, the most stable dispersed phase would eventually become the predominant one.

Determining which phase is continuous in a well-mixed liquid-liquid system is best accomplished visually. A sample is withdrawn from the tank during mixing and observed in a glass beaker. As the drops accumulate between two coalesced regions, it will be noted that the upper and lower surfaces of the discrete drops are not identical. The drops will lodge within the phase of which they are made because their density will match that particular phase.

The surface next to the matching phase is irregular. The other surface of the group of drops next to the opposite phase is quite flat because of this density difference. The composition of the drops is the same as the phase that presents the irregular surface.

XIV. APPLICATION OF PRINCIPLES

Principles of fluid mechanics and mixing developed in this chapter are of use both in chemical research for an understanding of the effect of mixing, and also to enable large-scale equipment to be applied successfully for industrial processing. Some applications to processing have been mentioned, but applications to operations where interfaces between phases are critical will be developed in the next chapter.

REFERENCES

1. F. A. Holland and F. S. Chapman, *Liquid Mixing and Processing in Stirred Tanks,* Reinhold, New York, 1966.
2. J. H. Rushton, *Mixing for the Chemical Industries,* American Institute of Chemical Engineers, Twenty-Second Institute Lecture, 1970. American Institute of Chemical Engineers, New York.
3. J. H. Rushton, How to make use of recent mixing developments, *Chem. Eng. Progr., 50*:587 (1954).
4. J. H. Rushton, and J. Y. Oldshue, Mixing of liquids, *Chem. Eng. Progr., Symp. Ser.,* 55, No. 25 (1959).
5. J. H. Rushton, E. W. Costich, and H. J. Everett. Power characteristics of mixing impellers, *Chem. Eng. Progr., 46*:395, 467 (1950).
6. V. W. Uhl and J. B. Gray, *Mixing,* Vols. 1 and 2, Academic, New York, 1966.

3.

HIGH-SPEED STIRRING IN INTERFACIAL SYNTHESIS

J. Henry Rushton[*]

School of Chemical Engineering
Purdue University
Lafayette, Indiana

I.	INTRODUCTION	38
II.	INTERFACIAL AREA	39
III.	PROPERTIES, FLOW AND AREA	41
IV.	USE OF EQUATIONS	44
V.	APPLICATIONS	46
VI.	FLOW-CONTROLLED SYSTEMS	47
VII.	HEAT EFFECTS	49
VIII.	GAS-LIQUID CONTACTING	50
IX.	POLYMERIZATIONS	54
X.	VISCOUS SYSTEMS	55
	REFERENCES	56

---

[*]Also technical advisor, Mixing Equipment Co., Rochester, New York.

I. INTRODUCTION

The term high-speed stirring is often heard in the organic laboratory. Equipment of the type of a Waring blender is used frequently in research laboratories to give high-speed stirring. Small discs, paddles, and propellers are the types of impellers used. Rotational speeds of 1,000 rpm and higher are spoken of as high speeds. The impellers are frequently only an inch or two in diameter. The power necessary to turn these impellers in low-viscosity liquids is small. The actual flow may be very small compared to the head or turbulence. High shear stresses are developed in the flowing stream, and small-scale turbulence will result.

Referring to the power and fluid mechanics relations of Chap. 2, it will be seen that high-speed stirring is synonymous with high head and turbulence. Thus, if high-speed stirring is essential, it really implies that turbulence is the desired factor and that flow is of minor importance. Frequently the organic chemist uses high-speed stirring in small laboratory vessels when it might not at all be necessary to apply the amount of turbulence resulting therefrom. In fact he may be using much more energy than would be required for optimum results. High-speed stirring devices are readily at hand and are frequently used for convenience. This is not to say that high-speed stirring is not useful or even essential for some reactions. Many reactions do require high turbulence compared to flow.

Another consideration is that of equipment size. When high-speed stirring is actually required it may become very difficult to reproduce the desired turbulence level in large-scale equipment. Hence, if high-speed stirring is used by the chemist it will be necessary at some time to determine exactly how much high-speed stirring is required. Frequently the laboratory high-speed devices will produce far more turbulence than is needed. In the pilot plant it may not be feasible to apply this level of turbulence, and at some point for industrial synthesis work it will probably be necessary to know the proper ratio of turbulence to flow for large-scale design.

3. High-Speed Stirring in Interfacial Synthesis

Another consideration is that the chemist may wish to be assured that fluid motion is sufficient and that a small change in fluid motion will not affect the overall kinetics of the reaction. For this reason he may deliberately want to overwhelm the system with vigorous agitation, hoping thereby to realize the rate effects of chemical kinetics alone. This, of course, is a proper objective in many cases, but at some point in the research program the chemist should be urged to decrease speed until there is actually a change in his result. He will thereby obtain more knowledge as to the effect of stirring on that system.

II. INTERFACIAL AREA

Interfacial area is important for chemical reactions for three reasons. First, the area provides points of contact for the molecules reacting. Second, the rate of mass transfer for material across films adjacent to the interface is directly proportional to the interfacial area and to the concentration gradient. This is true both for the reactants and for the products of reaction. Third, the rate of heat transfer across fluid films to and from an interface is directly proportional to the interfacial area and to the temperature gradient across the fluid layers.

Heat transfer and its rate are important to maintain a particular temperature desired, and heat-transfer rate changes can materially affect the kinetics of the chemical reaction involved.

The development of interfacial area is often a combined effect of the chemical species present and the forces of fluid motion due to the stirring. For a high interfacial tension the fluid motion forces can have a very great effect on the formation of interfacial area. Theoretically, fluid motion can be expressed in terms of force, length, and time. Recall that the properties of fluids may also be expressed in terms of force, length, and time. Since force and mass are related (force is equal to mass times acceleration) it is clear that mass may also be used with length and time to define fluid motion. Since physical properties and fluid motion are defined by the same physical units, it is clear that the fluid properties will interact with the physical motion of the fluids [1].

Fluid mechanics recognizes various forces; for example, the inertia force of Newton and the forces deriving from physical properties of fluids can be related. The Reynolds number of fluid mechanics is the dimensionless ratio of the inertia force to the force corresponding to the viscosity of a fluid. The Reynolds number may be formulated as a characteristic size or dimension (L) of a system (impeller diameter, D, for example), the velocity of flow (u) at some particular point in the system, the density (ρ) of the fluid, and the viscosity (μ) of the fluid. When these are arranged in the form $Lu\rho/\mu$ the result is a dimensionless number that expresses the effect of viscous forces.

In a similar way the ratio of Newton's inertia force to the force corresponding to interfacial tension results in a dimensionless group called the Weber number. It is expressed as Lu^2/σ where σ is the interfacial tension; these, of course, are in consistent units so that the number is dimensionless. When interfacial tension forces are active and when a fluid flow system is such that equilibrium can be obtained between the inertia force of Newton and the interfacial tension force, the effect of the Weber number on interfacial area between the two liquids involved is correlated. Ordinarily when equilibrium due to normal development of fluid motion can be obtained, the interfacial area can be directly related to both the stirring mechanism and the fluid properties. Note also that system size itself plays an important role, and as system size increases for a constant velocity it is clear that interfacial area will increase directly with size. Actually most data in the literature to date predict that interfacial area is a function of Weber number to approximately the one-third power.

Experimental results for stirring operations in vessels, reactors, and mixing tanks vary from this exponential relation. The variation appears to be due to the fact that when drops of one fluid are formed in another in the turbulent stream flowing from the mixing impeller

3. High-Speed Stirring in Interfacial Synthesis

toward the bottom or walls of the container, these drops strike the retaining surfaces. Some of them split to smaller drops, but many of them strike each other at the walls and form larger drops. Thus, the equilibrium drop size in a mixing vessel may not alone be due to fluid mechanics considerations of a simple-flowing system. Rather the normal development of equilibrium in a straight-flowing system is interrupted by the very sharp changes in direction of both the dispersed phase and the continuous phase at the walls of the vessel. The net result is that interfacial area has been found to be not easily related to the Weber group. It is also not easily related to the Reynolds group which takes viscosity into account, nor to another dimensionless group called the Froude group, u^2/Lg, which accounts for the ratio of inertial force to gravity force. The Froude group will also come into play when the difference in density of the dispersed and the continuous phase is considerable. Thus, the interfacial area is not easily related to the commonly used fluid-mechanics dimensionless groups, and recent research results relate the independent variables in more direct interaction relations [2]. The Reynolds and other groups are useful, however, in defining limits of viscous and other property and motion effects.

III. PROPERTIES, FLOW AND AREA

There has been a large number of publications relating to the development of liquid-liquid interfacial area in mixing vessels. These relations are summarized elsewhere [3, 4]. But agreement has not been reached as to the effect of interfacial tension, and the Weber group alone cannot account for the effect, although it indicates the effect qualitatively.

Very recent work, involving the dispersion of organic liquids in water and also of water in organic liquids, shows the very complicated dependence of interfacial area (a) on various fluid properties and on fluid-mechanics parameters in mixing tanks where flat-blade turbines are used [2]. A 17-term equation has been found to fit data over a wide range of physical properties and a threefold range in equipment size. This complicated relation can be reduced in complexity and is summarized as follows:

$$a = 0.186 \, \phi^\alpha \, \sigma^\beta \, \mu_d^\gamma \, \rho_c^\delta \, \rho_d^{-0.71} \, (P/V)^\varepsilon \, u^\eta \qquad (3.1)$$

where

$\alpha = 1.54 + 0.12 \ln(\phi) + 0.15 \ln(\mu_d)$

$\beta = 5.77 - 1.04 \ln(\sigma)$

$\gamma = 0.74 - 0.03 \ln(\mu_d)$

$\delta = 78.6 - 22.5 \ln(\sigma) - 6.93 \ln(\mu_d)$

$\varepsilon = 0.68 + 0.08 \ln(\phi) - 0.10 \ln(P/V)$

$\eta = 1.35 + 0.12 \ln(\mu_d) - 0.73 \ln(\mu)$

Although Eq. (3.1) is specific for a six-flat-blade turbine of the type shown in Fig. 2.9 it is clear from experience with such systems that the basic relations between the terms involved will be approximately correct for other types of mixing impellers. For different stirring devices the constant (0.186) will be different and the exponents on the parameters will vary. But one should not expect the variation to be very great. Furthermore, for any one chemical system in the research laboratory using impellers of the same type as used for stirring in an industrial operation, one can use ratios of the variables and predict with considerable confidence the effects of stirring.

Returning to Eq. (3.1), note that ϕ, the phase ratio (dispersed to continuous phase) has an exponent α; typical values of α are in the region of 0.66. The next variable is the interfacial tension, σ; the exponent for σ is β, and β will have values of around -0.45. Note also that for both α and β the values are likewise dependent upon the logarithm of the viscosity of the dispersed phase, μ_d. Continuing to the third variable, the viscosity of the dispersed phase, μ_d, the exponent γ may have a low value and is dependent upon the logarithm of itself. The next variable of importance is the density of the continuous phase, ρ_c. Its exponent, δ, has small values, and these values are functions also of σ and μ_d. The density of the dispersed phase, ρ_d, has a simple exponent of -0.71. The next

3. High-Speed Stirring in Interfacial Synthesis

term is a composite of impeller diameter, impeller speed, and continuous fluid density divided by the volume of fluid (which is proportional to the cube of vessel diameter, T). This variable is noted as power per unit volume, P/V, and is the power exerted by the impeller divided by the total volume of fluid. The exponent on this parameter is ε, which itself is dependent upon the logarithms of ϕ and P/V. Values of ε are small but significant. The final term u, the linear velocity of fluid leaving the impeller, has an exponent η whose value is relatively small but is dependent upon the logarithm of viscosity of the dispersed phase and its velocity.

The description of this equation indicates the complicated interactions between the fluid properties and fluid motion caused by the stirring device. The data are correlated extremely well by this relation. If one were to change the form P/V and u to terms of impeller size, container size, and impeller rotational speed, the equation could be set up to show a Weber number effect involving the interfacial tension and also a Reynolds number effect involving viscosity. However, if this is done, residual values of interfacial tension, viscosity, and density will be left over, showing that the dimensionless groups do not fully account for the interactions. Thus, little advantage is gained in forcing the concept of free equilibrium between the properties and theoretical fluid mechanics. Accordingly, it must be concluded that, although there is an effect of Weber number and the direction of this effect is correct, the Weber number itself does not represent the true and complete effect of interfacial tension. For this reason, and to simplify the concept of energy from stirring put into a volume of fluid, the term P/V has been used to modify Eq. (3.1). For turbulent conditions P/V is actually

$$P/V = \frac{4K_2 \rho_c N^3 D^5}{g_c \pi T^3} \tag{3.2}$$

Also, u is proportional to ND, or

$$u = K_5 ND \tag{3.3}$$

where K_5 is constant for a particular impeller shape.

IV. USE OF EQUATIONS

Without detracting seriously from the accuracy of Eq. (3.1) and the previous discussion it is possible to rearrange the equation wherein the phase ratio ϕ is held constant and the fluid properties also are held constant. Then, using D for impeller diameter, N for impeller speed, and T for container diameter, one can relate Eqs. (3.2) and (3.3) to interfacial area, A. Furthermore, when experimental work is done for a specific system and in a particular container geometry one can use the ratios of the area, speed, and diameter of both impeller and tank, and the net result is the following relation:

$$A_r = \frac{N_r^{1.11} D_r^{1.23}}{T_r^{0.18}} \tag{3.4}$$

where the subscript r indicates ratios.

This equation says that the ratio of interfacial area to impeller speed, impeller diameter, and tank diameter ratios is a relatively simple one. For example, if the container diameter is doubled (and liquid volume increases 8 times) and the impeller is operated at the same diameter and speed, the interfacial area per unit volume will be decreased from its original value to $1/2^{0.18}$, or 1/1.333, or 0.88 of the A in the smaller tank. By a similar calculation, if impeller speed is doubled and impeller size and tank size are held constant, the interfacial area will be increased by 116% to 2.16 times that for the lower speed. This effect is achieved by an 8-fold power increase (see Eq. 3.2). Likewise one could calculate the effect on interfacial area of changing the diameter of the impeller; and the interfacial-area ratio can also be predicted for changes in all three variables at the same time.

Referring to discussions in the preceding chapter involving flow and turbulence, one can take Eq. (3.4) and substitute ND^3 for flow and N^2D^2 for turbulence and determine a relation for these two dependent properties from the equation. However, it will be found

3. High-Speed Stirring in Interfacial Synthesis

that residual exponents will result. This is simply to say that the ratio between flow and head or flow and turbulence will not remain constant for a change in overall size, and in fact when the ratio of impeller diameter to tank diameter is changed the optimum ratio of flow to turbulence will not hold constant at constant power. Considerable emphasis has been given previously to the importance of head and flow. This is still true, but one can see from the interaction of size and fluid properties that the ratios of flow to head do not remain constant for significant changes in size of the system for dimensionally similar conditions. It must, however, be noted that optimum ratio of flow to head may vary only 10-15% as one increases the size of the system. However, systems that are primarily controlled by turbulence will be affected differently by a change in size of the impeller or the container than systems that are more dependent on flow.

Returning again to the concept of high-speed stirring and the formation of interface, suppose that the speed of the high-speed laboratory device is increased from 1,000 rpm to 2,000 rpm. The speed ratio of 2 raised to the power of 1.11 indicates that an interfacial area increase of approximately 116% will result. Thus a doubling of the impeller speed will more than double the interfacial area while the power required will be increased by the speed ratio cubed (which in this case is 8 times). Hence, we have developed 2.16 times the interfacial area at a cost of 8 times the amount of energy. Likewise there will be 8 times the amount of mechanical heat to be rejected or absorbed.

It is clear from this that increases in stirring rate can be used to make very significant increases in interfacial area; however, other problems may develop, and the cost in energy involved may be high. Obviously, there are many syntheses where the cost of mechanical power may be offset by the considerable benefits of increased interfacial area and the increased reaction rates resulting therefrom.

V. APPLICATIONS

A large number of reactions are well served by the increase in interfacial area caused by stirring or mixing. Alkylation units in the petroleum industry using sulfuric acid, hydrofluoric acid, and aluminum chloride are examples of these operations. Control of the system and reaction rates can be effected in major proportions by attention to the physical variables involved in mixing.

Another point that should be mentioned about relations such as those given by Eqs. (3.1) and (3.4) for interfacial-area development is that maxima occur if one plots interfacial area against D/T ratios for different values of power per unit volume (see Fig. 3.1). These maxima occur normally at D/T values of about 0.3, although sometimes values as low as 0.2 give maxima. A value of 0.3 means that the tank diameter is 3.3 times the impeller diameter. Thus, for a given value of power per unit volume, interfacial area per unit volume

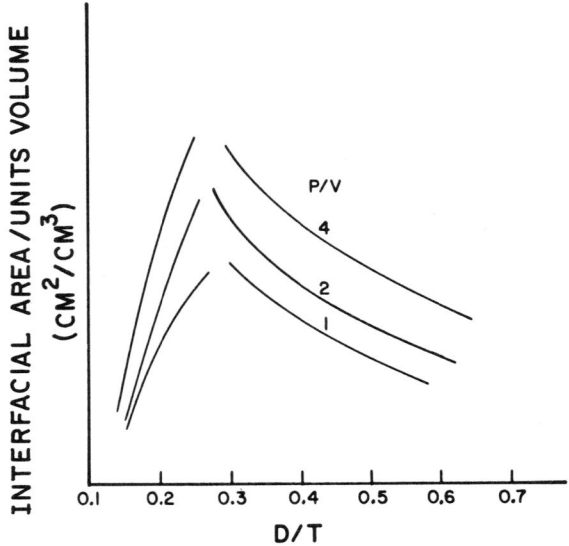

FIG. 3.1. Interfacial area as a function of impeller to tank diameter ratio, and of power per unit volume.

3. High-Speed Stirring in Interfacial Synthesis

will decrease as tank size increases above a D/T ratio of 0.3. A consequence of this relationship, which can of course be computed from Eqs. (3.1) or (3.4), is that more power per unit volume must be used to create the same interfacial area per unit volume when the tank or container size is increased. This, of course, is unfortunate. The chemist working with small containers has a relatively easy time in his efforts to create interfacial area per unit volume. It is much more difficult to reproduce his results economically in large-scale systems when they are dependent on interfacial area. Another way of saying this is that the power per unit volume must increase to attain the same interfacial area as one scales up to larger equipment.

VI. FLOW-CONTROLLED SYSTEMS

The emphasis thus far has been on two-phase immiscible liquid systems where the interfacial tensions are such that stirring and fluid motion can have a very considerable effect on drop size and area. However, there are numerous cases where dispersions can be formed almost by pouring one liquid into another. In such cases mixing and stirring may be of very minor importance and may be required simply to maintain a homogeneity of the entire system—particularly for reasons of mass transfer and possibly to prevent a settling of more dense phases.

Sometimes there are systems that can be maintained with high interfacial area with very small amounts of stirring. In these cases flow is probably the important requirement for mixing. However, note that flow is proportional to ND^3 whereas turbulence is proportional to N^2D^2; also note from Eq. (3.4) that the effect on interfacial area produced by the mixing impeller is almost proportional to ND, but not quite. Sometimes attempts are made to relate interfacial-area formation to impeller-tip speed. Impeller-tip speed is πDN; however, the constant π can be neglected for comparison. One can see that area varies closely with tip speed, although the actual effects of tip speed on area change as container size changes. The product ND is not sufficient to characterize interfacial-area formation.

When more flow is required than turbulence the relation between N and D will change. Actually most two-phase systems show interfacial-area changes as functions of N, D, and T, and the relations change as the container size changes.

One forceful example from data taken in five sizes of equipment from 6-in. diameter experimental units up to full-size equipment of 12 ft in diameter will serve to demonstrate the complexity of the effect of mixing. This example was for a polymerization where the polymer had considerable tendency toward sticking to itself. Large balls of polymer could be formed whenever too much turbulence was present. It was noted that tip speed had to be low to prevent this "balling." Further, as system size was increased the tip speed, which was critical for balling, actually decreased. When the data were considered in terms of flow rather than in terms of turbulence, and the data were applied in terms of ND^3 divided by T^3, the effect of tip speed on balling could be predicted. Thus, from a plot of flow per unit volume (ND^3/T^3) vs. container diameter (T) where the speed and diameter were at a critical condition with respect to the product quality (such as tackiness and balling), the relation was found to be $ND^{4.1}/T^3 = C$, where C was a constant whose value was 0.38. The tip speed for a given size of T could thus be calculated.

Another set of data for another reaction primarily dependent on flow, and for four tank sizes, was found to be $ND^{3.7}/T^3 = C = 0.45$.

Such relations are particularly useful for scale-up of laboratory data to large size. Note that the exponent of D (which indicates size, or scale) and the value of the constant C are unique for each chemical or polymerizing system.

This example shows clearly the effect of fluid motion on characteristics such as sticking and balling which could not otherwise be quantified by such an experimental program together with plant performance. It is clear that no simple relation such as tip speed can be effective as a control device to operate safely and properly with different sizes of equipment.

3. High-Speed Stirring in Interfacial Synthesis

Although the research chemist may not have this in mind when he starts a synthesis involving interfacial area, it is well for him to understand what will happen when he or others change equipment size and also that there are accountable reasons for many effects that at first glance are very peculiar. Frequently there are operations wherein the products of reaction have a tendency to stick on metal or other materials of the equipment; often glass containers and glass impellers are found to give less trouble from the standpoint of sticking to reactor surfaces.

Of course sticking may become prevalent on cooling surfaces, but the majority of reactions do require cooling-type heat transfer and methods must be recognized to be able to handle such systems. Glass equipment in the laboratory is easy to handle, but it is not easy to obtain good, sturdy equipment made of glass for large-scale operations. Effort, therefore, is made to find some form of stainless steel or other metal which can be used economically. For many organic syntheses it will be noticed that impeller speed may have a significant effect on the tendency of material to stick; often such effects are caused by too high an impeller speed. Much lower speeds may be required. The dilemma which then must be faced is to use an impeller to create interfacial area, thus giving high shear stress in the fluid flow, when lower shear stresses are required to reduce the tendency to stickiness or tackiness. Furthermore, heat transfer rate is dependent on flow motion, but fortunately it is more dependent upon flow than upon turbulence, and some concessions may have to be made to interfacial area to achieve better heat transfer.

VII. HEAT EFFECTS

A further complication when heat transfer is involved is the fact that as the system becomes larger the volume is a function of linear size (e.g., diameter) cubed, whereas heat transfer areas are a function of linear size squared (this of course, is exactly true for geometrically similar systems). Therefore, if there is the same heat given off by reaction per unit volume of fluid, then in larger geometrically similar systems the heat transfer requirement per unit area of cooling surface becomes greater and exactly proportional to the linear size increase. Proper balance of mixing conditions must be used to achieve economic operation, but in some cases if too large a system size is used one must resort to an auto-refrigeration technique whereby low pressure can be used to distill off some of the continuous phase, thereby removing heat from the reaction as latent heat.

VIII. GAS-LIQUID CONTACTING

Interfacial contacting of gas and liquid by means of stirring is frequently of use for a wide variety of operations; for example, hydrogenation of fatty oils is commonly practiced, and oxidation for production of penicillin and other such processes whereby air or air enriched with oxygen is widely used. Such operations depend upon interfacial area, and mass and heat transfers are involved, just as for cases of liquid-liquid operations mentioned above. The effect of mixing variables and fluid mechanics is quite pronounced due to large differences in specific gravities between gas and liquid. High interfacial tension is involved and considerable energy is required to produce the interfacial area in the form of bubbles and to hold the gas under the liquid surface for periods of time so that mass transfer can be accomplished.

Two important mechanisms are involved. Area must be formed and materials must be transferred to the interface. Stripping and absorption are common operations for this purpose; often complications arise because reactions may take place in the liquid phase rather than at the interface, and the products of the reaction must move out into the gaseous phase. For example, in biological operations in deep bed fermenters mixers are used to aid the transfer of oxygen from air to the liquid and to aid in the transport of the dissolved oxygen to the growing material. Then, the product carbon dioxide must pass through the liquid and the interface to the gas bubble. If the effect of turbulence and mixer action can be assessed for

3. High-Speed Stirring in Interfacial Synthesis

each of these mechanisms it is possible to predict performance. Thus, the research chemist should vary the speed and size of impellers to determine the mixing parameters and their effects on interfacial area.

There are some data in the literature on bubble sizes which can be achieved with mixers [3], but the common approach to this problem is to relate performance to mass transfer coefficients and show how these coefficients vary with gas flow rate, mixer size, and speed [5]. Absorption coefficients are normally shown as a function of mixer power and gas rate. An illustration for absorption coefficients is shown in Fig. 3.2. Here, the mass transfer coefficient is plotted against power per unit volume for different air-flow rates. It will be noted that the superficial air flow rate has a considerable effect on the transfer coefficient for constant power per unit volume. For any one air flow rate the power requirement increases rapidly to effect an increase in the mass transfer coefficient; this

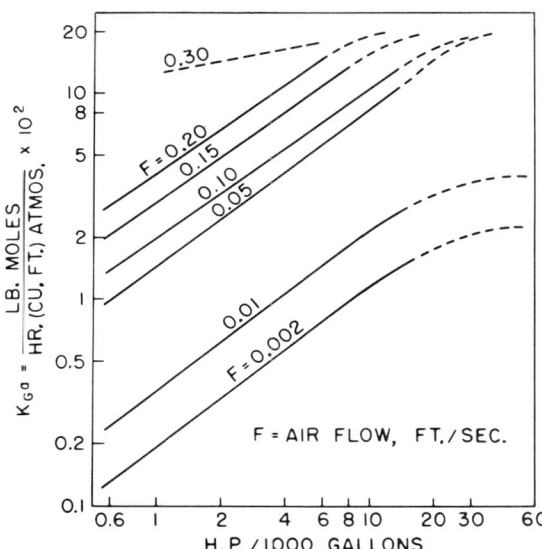

FIG. 3.2. Absorption coefficients for single turbines.

mass transfer coefficient is essentially directly proportional to interfacial area.

Again the optimum ratios of D/T depend upon mixer power and gas flow rate at a given level of power input. Small differences in impeller positions and liquid level to tank-diameter ratio do not affect absorption coefficients greatly. Ordinarily liquid depth is at least equal to tank diameter but is often greater. Multiple impellers, although frequently used, may or may not give results superior to those of single impellers. The ratio of liquid depth to tank diameter for most effective gas-liquid contacting is critical and depends upon the volume of liquid and the rate of gas flow.

Total power required in gas-liquid contacting involves not only the power imposed by the rotating impeller but also the power required to compress and supply air. When air quantities increase, air-pressure requirements usually increase and the power in the air stream increases. This in turn usually requires higher speeds for the impellers; but since they operate in lower density surroundings, the mixing power required may actually decrease. A point will be reached where decreasing mixing power and increasing compressor power will allow selection of an optimum.

Superficial gas velocity is the volume of gas divided by the cross-sectional area of the tank. This velocity will increase as tank diameter decreases for the same quantity of flow, and liquid level would increase for the same total liquid in contact. An increase in tank height requires a longer shaft for the mixer, and there is an economic limit to this length in most cases. Accordingly, tank proportions are fixed by proper balance between these factors.

Gas should be introduced either with a center inlet directly below the impeller or with a sparge ring to bring the gas to the periphery of the impeller. Any center inlet should be carefully centered below the impeller so that gas distribution will be uniform. Gas in the liquid passing through an impeller will result in an apparent low density, and from Eq. (2.2) it is clear that the power

3. High-Speed Stirring in Interfacial Synthesis

required to drive a mixer at a given speed when gas is present will be less than if no gas is present. Thus, for a given speed, the power required to drive the mixing impeller will decrease as gas is fed, and for proper design power requirements must be based on gassed conditions.

The design basis for mixing power required for a particular overall result is that which occurs during gassing. The difference in power required to turn the impeller gassed and ungassed may be very large, by as much as a ratio of 1 to 5. Thus, mixers must be protected by overload relay, so that if gas flow suddenly stops, the mixer motor will not be overloaded. Furthermore, it is common to use two-speed motors to drive impellers so that proper power may be applied during gassing, and also as desired during times when mixing is to take place without gas flow. The horsepower ratio between ungassed and gassed operations is a function of the rate of gas flow and of the manner of introduction of gas. Although at high power rates better gas distribution and bubble refinement can be achieved by a center input, the choice between using a center inlet for a gas or a ring sparger at the edge of the impeller is one of economics in the supply of motors, mixer assemblies, and methods of operating at different speeds required by the effect of gas and ungassed power requirements.

Gas absorption or mass transfer coefficient data should be obtained so that proper scale-up may be made. Such coefficients put in the form of the Sherwood number can be related to power. The absorption coefficient is a function of mixer size and speed, and of gas flow rate. Absorption coefficients can be predicted for large-scale operation in a manner exactly similar to that for scaling up heat transfer and other mass transfer liquid-liquid operations.

Sometimes it is very inconvenient or difficult to obtain absorption coefficients. It is then sufficient to know the time of contact necessary between gas and liquid. Data on holdup and contact time can easily be obtained and related to impeller speed and gas flow rate. Scale-up can be made to produce equal amounts of holdup per unit volume or contact time per unit volume.

IX. POLYMERIZATIONS

There are a number of techniques for polymerization, many of which involve two phases at the start. The common methods are emulsion, suspension, and solution polymerizations, in addition to normal bulk polymerizations.

Suspensions and emulsions are produced so that polymerization can take place in the dispersed phase, although in many cases there is phase inversion. This is of common occurrence with many styrene copolymers. In any case immiscible liquids are either formed mechanically or develop from the reaction and then modified by the mechanical stirring. From the previous discussion it is clear that the "level" of agitation is of extreme importance, and by level of agitation is meant the amount of turbulence and flow of material, and this in turn is frequently referred to as power input per unit volume.

There are frequent references in the literature and wideapread ideas involving power per unit volume. Power to produce flow and turbulence can be achieved by a small impeller at high speed or a large impeller at low speed. But if both these impellers are run at the same power the difference in flow and turbulence in each case would be significant in the formation of interfacial area, particularly as applied to polymerizations involving emulsions and suspensions. Accordingly, one should *never* expect to use the same power per unit volume in different-size equipment to produce the same result. Almost *never* is it proper to scale-up systems on the basis of equal power per unit volume in heat transfer operations. One must always use more power per unit volume on a scale-up to achieve equal interfacial areas per unit volume. However, power per unit volume is a useful concept if properly applied. If the chemist can determine the increase in power per unit volume required for larger equipment he will have taken a big step toward commercialization of a laboratory polymerization reaction.

Many interfacial polymerizations involve reactants like carboxylic acid chlorides where activation energies are low—on the order of 3 kcal/mole and lower. Here the reaction rates are diffusion controlled. For such operations the mixing requirements are easily met by using relatively large impeller to tank diameters. This implies a mixing requirement favored by high flow and low turbulence. In such cases the scale-up to plant-size equipment can be done easily and economically.

X. VISCOUS SYSTEMS

There are many polymerizations (especially of suspension type) where the fluids become very viscous and behave as non-Newtonian fluids. The relations between power, speed, size, and fluid properties can be treated as for Newtonian fluids as shown in Eqs. (2.2) and (2.3), provided that the values for constant K_2 and K_3 are determined for the non-Newtonian properties [6].

It is much more difficult to create interfacial area in highly viscous liquid systems since turbulence will not persist when viscosity is high, and most polymerizations are such that the polymerizing material becomes highly viscous. In suspension polymerizations the continuous phase may be kept at a low viscosity and the more viscous polymer handled as the dispersed phase. When the viscosity of the continuous phase is high and the Reynolds number for mixing is below about 1,000, the fluid motion conditions are probably entering the viscous range [7, 8]. In this case Eq. (2.3) gives the relation for viscosity, speed, size, and power. Turbulence will be initiated close to the impeller and at the edges of the discharged stream, but the turbulence cannot persist and is damped out quickly. Accordingly, to initiate interfacial area in highly viscous liquids it is necessary to use rotating-bar-type impellers and usually to locate these very close to heating or cooling surfaces so that high shear stresses can be developed between the fluid adjacent to the impeller and the fluid close by. Such special mixing devices are not standard items and must be made to order; however, there are some commercial devices for viscous fluids where close clearance is held between the rotating impeller and the surrounding walls. Such devices are essential for certain very high viscosity uses. For example, to disperse a high-viscosity material or solid into a continuous phase of very high viscosity may require very close clearances for rotating bodies. These rotating impellers serve to attenuate the dispersed phase, stretching it past its elastic limit until a new interface is formed which cannot return to its previous condition.

For lower viscosity continuous-phase materials colloid mills are used to generate interfacial area; these are usually devices with one rotating element passing close to a fixed element, and continuous flow of two phases is caused to pass between.

REFERENCES

1. J. H. Rushton, Applications of fluid mechanics and similitude to scale-up problems, Chem. Eng. Progr., 48:33, 95 (1952).

2. C. M. McLaughlin, The Formation and Measurement of Liquid-Liquid Interfacial Areas in a Mixing Tank, Ph.D. Thesis, Chemical Engineering, Purdue University, August, 1972.

3. P. H. Calderbank, Part I: The interfacial area in gas-liquid contacting with mechanical agitation, Trans. Inst. Chem. Eng. (London), 36:443 (1958).

4. W. A. Rodger, V. G. Trice, and J. H. Rushton, Effect of fluid motion on interfacial area of dispersions, Chem. Eng. Progr., 52:515 (1956).

5. J. H. Rushton and J. Y. Oldshue, Mixing of liquids, Chem. Eng. Progr., Symp. Ser., 55, No. 25 (1959).

6. P. H. Calderbank and M. B. Moo-Young, The power characteristics of agitators for the mixing of newtonian and non-newtonian fluids, Trans. Inst. Chem. Eng. (London), 39:22 (1961).

7. A. B. Metzner, R. H. Feeks, H. L. Ramos, R. E. Otto, and J. D. Tuthill. Agitation of viscous newtonian and non-newtonian fluids, A.I.Ch.E., 7:3 (1961).

8. J. H. Rushton, E. W. Costich, and H. J. Everett, Power characteristics of mixing impellers, Chem. Eng. Progr., 46:395, 467 (1950)

4.

APPLICATION OF TWO-PHASE SYNTHESIS IN ORGANIC CHEMISTRY

David J. Goldsmith
Department of Chemistry
Emory University
Atlanta, Georgia

I.	INTRODUCTION	57
II.	REACTION IN THE AQUEOUS PHASE	60
	A. Two-Phase Oxidation	60
	B. Two-Phase Dehydration	66
III.	PHASE-TRANSFER CATALYSIS	68
	REFERENCES	73

I. INTRODUCTION

Experimental procedures commonly used in synthetic organic chemistry may be divided into two broad classes. These classes are simply the familiar techniques of carrying out a reaction either in homogeneous solution or in heterogeneous mixture. Depending upon the particular process, each of these techniques has advantages. For example, the acyloin reaction [1], in which reductive cyclization of a long-chain diester occurs, is carried out most commonly in toluene solution in the presence of metallic sodium. The advantage of heterogeneity in this case results from the presumed coordination of both reactive sites of the diester on the *surface* of the sodium metal. The carbon alkylation of certain types of ambident anions is also best carried out heterogeneously. The work of Kornblum and co-workers [2] has shown that phenolate salts undergo significant carbon alkylation, e.g., [1] to [2], when the reaction is carried out in media in which

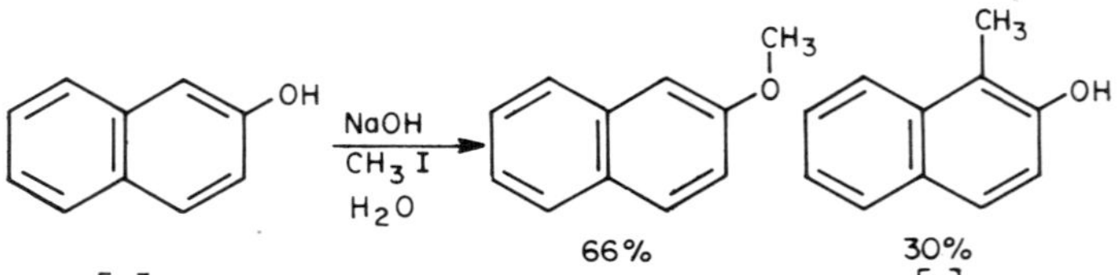

[diagram: [1] 2-naphthol + NaOH / CH₃I / H₂O → 2-methoxynaphthalene (66%) + 1-methyl-2-naphthol [2] (30%)]

the salts are insoluble. Thallium salts of β-dicarbonyl compounds also give very high percentages of C-alkylation when caused to react as solids, in contrast to the results obtained with the corresponding soluble alkali metal salts [3]. Heterogeneity is also advantageous in many instances where a nonsoluble catalyst rather than a nonsoluble reactant is employed. The most familiar of such laboratory procedures is undoubtedly catalytic hydrogenation in which the metal catalyst may be simply removed by filtration at the completion of the reaction [4]. Acid-catalyzed processes may often be carried out in the presence of an insoluble catalyst, such as a sulfonic acid resin [5], which again is readily removed from the reaction mixture without the necessity for an elaborate work-up procedure. The use of heterogeneous catalysts in the chemical industry is of course well established.

The quest for homogeneous reaction conditions is probably the most commonly followed approach in the development of organic synthetic procedures. Homogeneous reaction conditions are clearly simpler to control and comprehend than are the usual heterogeneous ones. Most obviously, the concentrations of various reactants and catalysts can be simply and accurately measured in homogeneous solution. More importantly, certain types of processes have been limited in their use prior to the development of suitable solvents. One particularly important example in organic synthesis is enolate alkylation. The ability of the synthetic organic chemist to control the position and frequency of alkylation of a selectively produced enolate ion is in large part due to the development and use of a variety of

4. Two-Phase Synthesis in Organic Chemistry

dipolar aprotic solvents [6]. These solvents, such as dimethoxyethane, DMSO, diglyme, THF, HMPA, etc., not only dissolve metal enolates but may, in appropriate proportion, be used to control the degree of aggregation of such salts.

One experimental technique for organic reactions which has seen little application is two-phase synthesis. The specific two-phase systems to be discussed here will be aqueous-organic liquids, in which the interaction of the reacting species may occur either in the bulk organic phase, the bulk aqueous phase, or at the interface of the two solvents. The question of where reaction takes place remains unsettled for many two-phase organic reactions. A few of the systems in which this problem has been explored have been reviewed by Menger [7]. Thus, for example, Bell [8] has shown for the permanganate oxidation of benzoyl-o-toluidine in water-benzene that reaction occurs at the interface. In contrast, the oxidation of tetrachlorohydroquine by ceric ion in carbon tetrachloride, 2-octanone-water, appears from the work of Mansoori and Madden [9] to occur in the aqueous phase.

In an elegant and definitive study of the two-phase (water-heptane) imidazole catalyzed hydrolysis of p-nitrophenyl laurate, Menger [10] has shown that this process takes place at the solvent-solvent boundary region. In another system more closely related to those to be discussed here, Menger [7] obtained evidence for the interfacial oxidation of borneol and isoborneol to camphor. The processes that will now be considered have not been fully explored with regard to the actual locus of reaction but appear from the available evidence to be, first, those in which an organic substrate is extracted into an aqueous phase where it undergoes structural change and the resulting product is then extracted into the organic phase, and, second, those in which a reactant or catalyst is transferred in small concentration from the bulk aqueous phase to the bulk organic phase to initiate a particular synthetic organic process ("phase-transfer catalysis").

II. REACTION IN THE AQUEOUS PHASE

A. Two-Phase Oxidation

One of the most commonly employed organic synthetic procedures is the conversion of a secondary alcohol to a ketone by Cr^{VI} reagents. Among the many variations [11] of this method

$$\underset{H}{\overset{OH}{>\!\!C\!\!<}} \quad \xrightarrow{Cr^{VI}} \quad >\!\!C=O$$

are the uses of sodium dichromate in aqueous sulfuric acid [12], anhydrous acetic acid [13], acetic acid-benzene [14], and acetic acid-acetic anhydride [15] and chromium trioxide in aqueous acetone-sulfuric acid (Jones reagent) [16], pyridine (Sarett reagent) [17], pyridine-water (Cornforth reagent) [18], and pyridine-methylene chloride (Collins reagent) [19]. One problem encountered with many of these methods, particularly those employing Cr^{VI} reagents in acid solution, is the possibility of further oxidation of the ketone product or of oxidation of other functional groups in either the substrate alcohol or the product [20, 21]. In addition, if the product of oxidation has an asymmetric center alpha to the ketone function, the strong acid often employed may cause epimerization of that center through enol formation [22].

A number of specific two-phase methods have been reported which avoid these difficulties. The "Organic Synthesis" procedure for the oxidation of dihydrocholesterol [3] to cholestanone [4] employs

[3] → [4]

4. Two-Phase Synthesis in Organic Chemistry

sodium dichromate in a mixed solvent system of water-acetic acid-benzene containing sulfuric acid [23]. The preparation of 2-cyclopentene-1, 4-dione [6] by oxidation of the corresponding diol [5] is also an "Organic Synthesis" [24] procedure. In this case the organic phase is methylene chloride. Another two-phase oxidation, again using benzene, is the Organic Synthesis [25] preparation of 2-methylcyclohexanone [8] from 2-methylcyclohexanol [7]. In all these cases excessive oxidation is apparently limited

[5]   →   [6]

[7]   →   [8]

by use of the two phase procedure. The organic solvent removes the ketonic product in each example from extended contact with the Cr^{VI} oxidizing agent.

One of the first examples of the use of two-phase oxidation to avoid epimerization of an enolizable dissymmetric ketone was reported by Johnson et al. [26]. A mixture of the decalin diols [9] when

[9]   →   [10]   +   [11]

treated with sodium dichromate in sulfuric acid-water-acetic acid-benzene afforded a 70% yield of the cis- and trans-diketones, [10] and [11]. The cis-isomer had not been previously reported, since under homogeneous acidic conditions it is isomerized to the trans compound [11]. The same procedure has also been employed by Johnson [27] for the preparation of cis- and trans-α-decalones, [12] and [13].

[12]   [13]

An extensive study of two-phase oxidation has been carried out by Brown et al. [22, 28]. These authors found that the oxidation of secondary alcohols with sodium dichromate in a system of diethyl ether-water-sulfuric acid was rapid, clean and relatively simple to carry out. They observed that a typical alcohol, cyclohexanol, rapidly disappeared from the organic phase (ether) followed by the appearance "at a moderate rate" in the same phase of cyclohexanone. Since the oxidation of alcohols by Cr^{VI} compounds is known [29] to occur by initial formation of a chromate ester, presumably the formation of that species (at the interface?) renders the substrate water soluble. Following elimination of the alcohol α-proton and the reduced chromium species [22] in the bulk aqueous phase, the resulting ketone product is then extracted again into ether.

This procedure has been shown [28] to be far superior to both homogeneous and two-phase oxidation with other organic solvents. Water-miscible solvents like tetrahydrofuran and diglyme consume oxidizing agent themselves or, as in the case of acetone (Jones oxidation), lead to some epimerization of dissymmetric ketones. Thus, for example, the two-phase oxidations of ℓ-menthol [14] and isopinocampheol [15] in ether-water mixture lead to yields of 97% and 94%, respectively, of the corresponding ketones, [16] and [17]. In both cases only trace amounts of the products of isomerization

4. Two-Phase Synthesis in Organic Chemistry

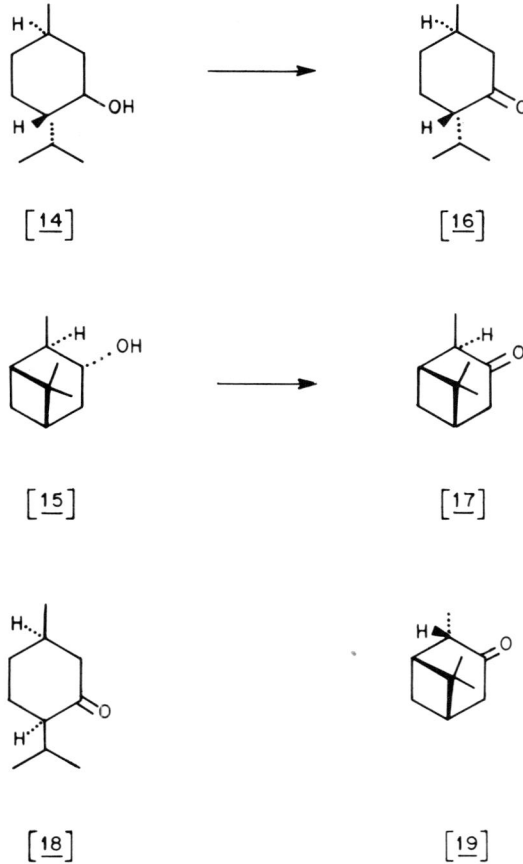

[18] and [19] could be detected. In contrast, the employment of chromic acid in acetone, aqueous chromic acid, or chromic acid in 90% acetic acid-water gave reduced yields and increased amounts of isomerized material.

Water-immiscible oxygen-containing solvents such as ethyl acetate, diisopropyl ether and diethyl ketone were also shown to consume oxidizing agent. Finally, water-immiscible solvents such as benzene and pentane though immune to attack by the oxidizing agent produced emulsions. These were sufficiently stable to make product isolation difficult.

The Brown procedure has been successfully applied to the oxidation of a wide variety of secondary alcohols. Indeed the list of such examples is so long [30] that no attempt will be made here to exemplify them all. The following examples will serve to show the versatility of this method.

Brown and co-workers have oxidized a variety of terpenoid alcohols to ketones unstable with respect to their epimers. In addition to the examples already cited, (-)-2-isocaranol [20] has been converted to (-)-2-isocaranone [21] [31]. The latter is

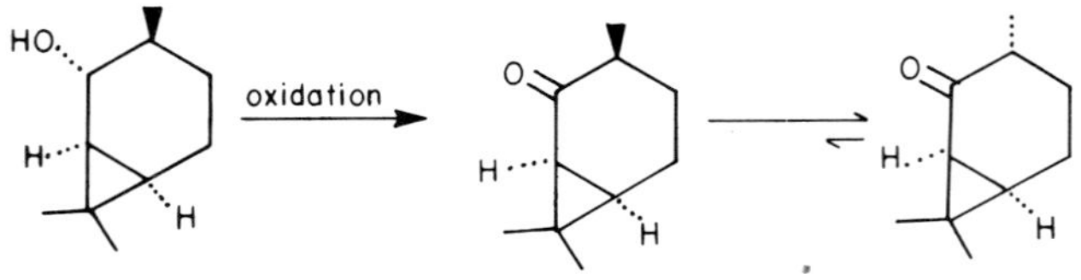

[20]   [21]   [22]

unstable with respect to its isomer (-)-2-caranone [22]. Similarly a mixture of 3-isothujopsanol [23] and 3-neoisothujopsanol [24] has been converted to 3-isothujopsanone [25] [32].

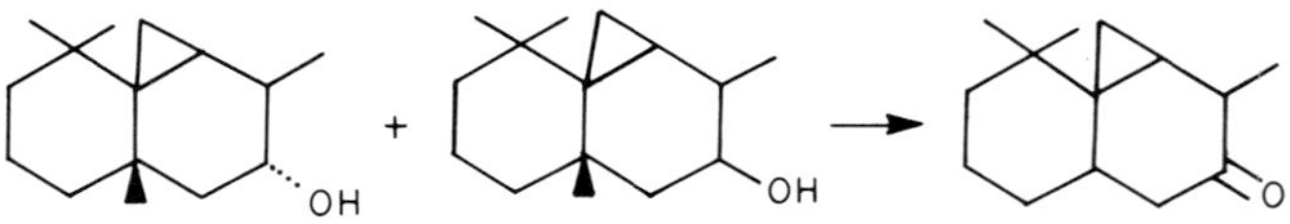

[23]   [24]   [25]

During the course of a study of epoxyolefin cyclization [33] we had occasion to prepare alcohols [26] and [27] by hydroboration of the unsaturated acetate [28]. The steric course of this hydroboration reaction was determined by oxidation of the alcohol product by the Brown method followed by pyrolytic elimination of the acetate. In this case the ketone product contains principally the more stable of the two epimers [29]. The oxidation reaction was not responsible for the predominance of the more stable product, however, since the

4. Two-Phase Synthesis in Organic Chemistry

[26]   [27]   [28]

[29]

percentages of the two ketones in the mixture were significantly different from the actual equilibrium values.

In other applications in the terpenoid field, the tricyclic alcohol [30], a system prone to acid-catalyzed rearrangement, has been converted to ketone [31] [34]; and hydroboration-oxidation of [32] followed by two-phase oxidation produces [33] without epimerization [35].

[30]   [31]

[32]   [33]

The Brown two-phase procedure has also been used for selective oxidation of a steroidal diol [34] to a ketol [35] [36], and is

[34]   [35]

reported [37] to be an excellent method for the oxidation of allylic alcohols to α β-unsaturated ketones, e.g., [36] to [37].

[36]   [37]

B. *Two-Phase Dehydration*

The second two-phase synthetic to be discussed here is the dehydration of alcohols in aqueous sulfuric acid-hydrocarbon mixtures. This method though perhaps widely used has been little commented upon in the literature and does not appear to have been systematically studied. The most interesting examples of two-phase dehydration in the last several years have come from the work of McMurry. The method employing hexane-50% aqueous sulfuric acid has been used, inter alia, in a total synthesis of sativene [38], for the dehydration of alcohol [38] to [39], in the total synthesis of longifolene [39] for conversion of [40] to [41], and for the preparation of enones [42] and [43], intermediates in a total synthesis of copacamphene [40].

The presumption for these reactions is that they occur in the bulk aqueous phase. Either protonation of the alcohol or formation of a sulfate ester would suffice to render the substrate water soluble. Following elimination, the resulting olefin would reenter the organic

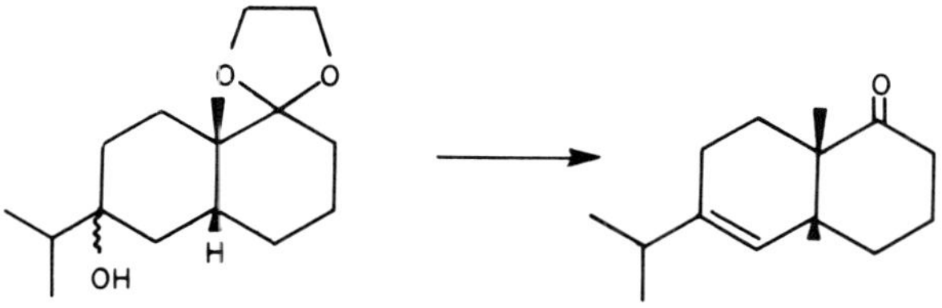

4. Two-Phase Synthesis in Organic Chemistry

[38]

[39]

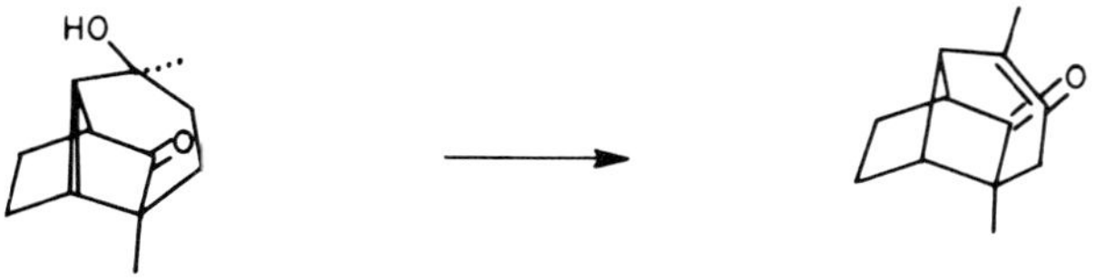

[40]

[41]

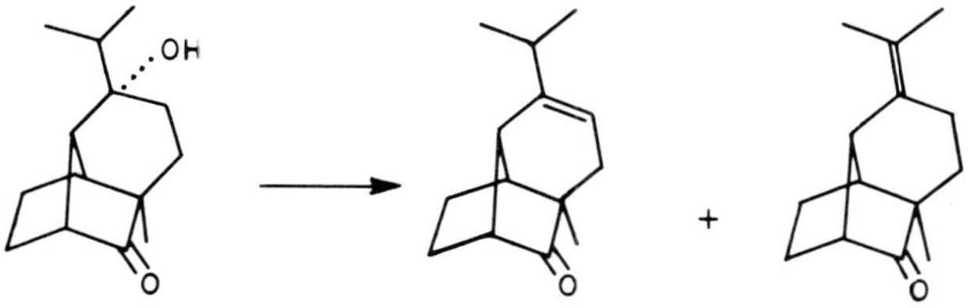

[42]

[43]

solvent. Little experimental evidence appears to be available on this point. Support for this concept, however, comes from the work of Ratcliffe and Heathcock [34] on the acid-catalyzed rearrangement of tricyclo[4.4.0.0^{2,7}]decan-3-ols. Alcohol [44] upon treatment with pentane-50% aqueous sulfuric acid is converted to the bicyclic olefin [45]. In contrast, the consanguineous olefins [46] and [47] are *unaffected* by the same reaction conditions. Heathcock states as his conclusion from this result, "apparently these olefins are not sufficiently protonated by sulfuric acid at this concentration to be drawn into the aqueous phase where reaction must occur." His suggestion is supported by the fact that treatment of [46] and [47] with pentane-70% sulfuric acid produces a mixture of polymeric products.

[44] → [45]

[46]   [47]

III. PHASE-TRANSFER CATALYSIS

A two-phase technique that shows great promise for application to a variety of synthetic reactions is "phase-transfer catalysis." As defined by Starks [41] this process is one in which "reaction is brought about by the use of small quantities of an agent which transfers one reactant across interface into the other phase so that reaction can proceed. The phase-transfer agent is not consumed but performs the transport function repeatedly." The usual phase-transfer agents are either tetraalkylammonium or tetraalkylphosphonium salts, the reactant being transferred is invariably an anionic species. The processes may be represented by the following equations:

$$X^- + Q^+Y^- \rightleftarrows Q^+X^- + Y^- \quad \text{(aqueous phase)}$$
$$Q^+X^- (aq) \rightarrow Q^+X^- (org)$$
$$Q^+X^- + \text{reactants } Z \rightarrow \text{products } X + Q^+Z^- \quad \text{(organic phase)}$$
$$Q^+Z^- (org) \rightarrow Q^+Z^- (aq)$$
$$Q^+Z^- + X^- \rightleftarrows Q^+X^- + Z^- \quad \text{(aqueous phase)}$$

where X^- may be, for example, OH^-, CN^-, or Hal^-, and Q^+Y^- is the quaternary salt (if Y^- and X^- are the same anion the first equilibrium

4. Two-Phase Synthesis in Organic Chemistry

does not apply). Reactants Z may be a single compound with a displaceable group, i.e., an alkyl halide or tosylate, or two organic species one of which reacts to liberate an anion, Z^-. Products X may represent a single product containing X, e.g., an alkylcyanide, or several products of which one will incorporate the X group, e.g., water, when X^- is hydroxide ion.

Among the processes that have been carried out by this technique are halocarbene formation, carbanion alkylation, and nucleophilic displacement.

The formation of dichlorocyclopropanes by the addition of dichlorocarbene to olefins is normally carried out under strictly anhydrous conditions [42]. In aqueous solution the conversion of chloroform to dichlorocarbene is blocked by hydrolysis of the intermediate $^-CCl_3$ to carbon monoxide and formate ion. Makosza, a pioneer in the field, found, however, that good yields of cyclopropanes could be obtained if a two-phase process was employed [43]. When a mixture of chloroform and styrene [48] was treated with a 50% aqueous sodium hydroxide solution containing triethylbenzylammonium chloride (TEBA), the cyclopropane [49] was produced in 80% yield. The formation of the intermediate carbene [50] occurs in the organic phase and results

$$\text{[48]} + :CHCl_2 \longrightarrow \text{[49]}$$

$$Et_3\overset{+}{N}CH_2\phi \ OH^- + CHCl_3 \longrightarrow CHCl_3$$

$$CHCl_3 \longrightarrow :CHCl_2 + Cl^-$$

[50]

from the migration of triethylbenzylammonium hydroxide across the interface. Since the concentration of olefin in the organic phase is high and the concentration of water in that phase is extremely

cyclohexene + CHCl₃ →(TEBA) dichloronorcarane [51]

n-C₄H₉-O-CH=CH₂ + CHCl₃ →(TEBA) n-C₄H₉O-CH(Cl)-CH₂(Cl) [52]

low, the carbene reacts preferentially with the olefin to yield a cyclopropane. Makosza has carried out this process with a variety of other olefins. For example, cyclohexene affords dichloronorcarane [51] in 72% yield, and butyl vinyl ether gives [52] in 71% yield under these conditions. The process has also been extended to the formation and addition of a thiocarbene [44]. Dichloromethylphenyl sulfide [53] reacts with styrene in the same two-phase system to afford [54].

PhSCH₂Cl [53] + styrene →(OH⁻, TEBA) [54]

Joshi [45] has also studied the phase-transfer catalytic reactions of carbenes and found that cetyltrimethylammonium chloride was an even more efficient catalyst than the triethylbenzyl compound. The yield of dichloronorcarane [51] from cyclohexene was increased to 98% by the use of the former salt. Relative rates of olefin reaction with dichlorocarbene in phase-transfer catalysis has been studied by Starks [41]. He found, not unexpectedly, since carbenes are electrophilic species, that highly substituted alkenes react faster than terminal olefins.

Makosza's two-phase conditions for dichlorocarbene formation have also been employed for other carbene reactions. Yoshida and co-workers have found (using benzene as a co-solvent) a highly efficient insertion [46] of dichlorocarbene into adamantane to form [55],

4. Two-Phase Synthesis in Organic Chemistry

[structure] → CHCl₃ / TEBA / ⁻OH → [structure with H, CHCl₂]

[55]

and a mild basic process for the conversion of alcohols to chlorides [47], e.g., [56] to [57]. The latter reaction occurs principally with retention of configuration indication an $S_N i$ mechanism for the substitution process.

[56] (norbornyl-OH) → CHCl₃ / TEBA / OH⁻ → [57] (norbornyl-Cl)

Carbon alkylation has also been found to be an efficient and easy process when a two-phase system is employed. Makosza first reported in 1965 [48, 49] the ethylation of phenylacetonitrile [58] to yield [59] employing a variety of alkyl ammonium salts. These workers studied the effect of variations in base concentration, halide concentration and alkylating agent structure. High yields (though often low conversion) were obtained with such representative alkyl halides as propylbromide, butyl bromide, allyl chloride and α-phenylethyl chloride [50, 51, 52]. They have extended their method, inter alia, to the alkylation of indene [53], e.g., [60] to [61], the

[58] PhCH₂CN → CH₃CH₂Cl → [59] PhCH(CH₂CH₃)CN

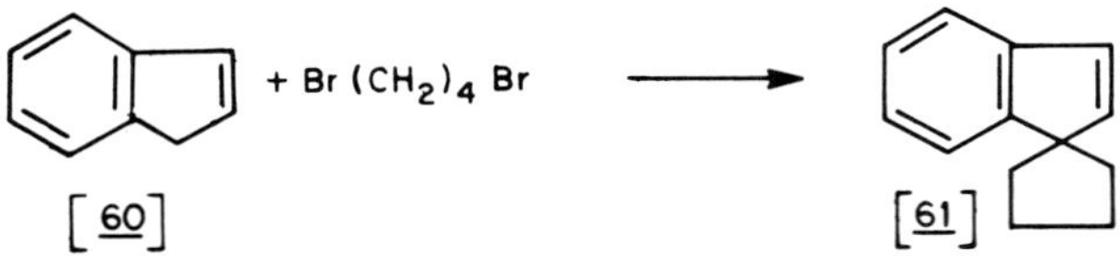

[60] + Br(CH$_2$)$_4$Br ⟶ [61]

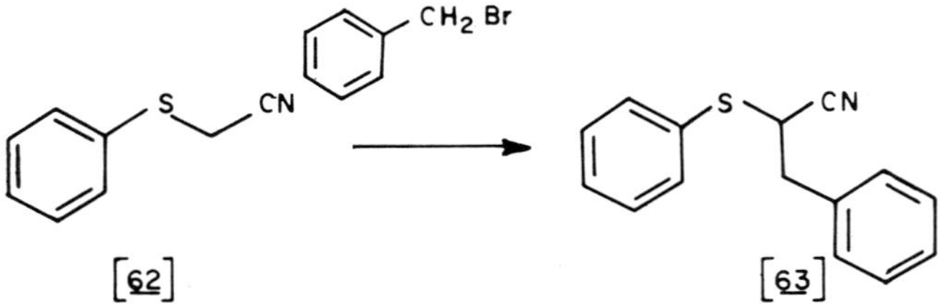

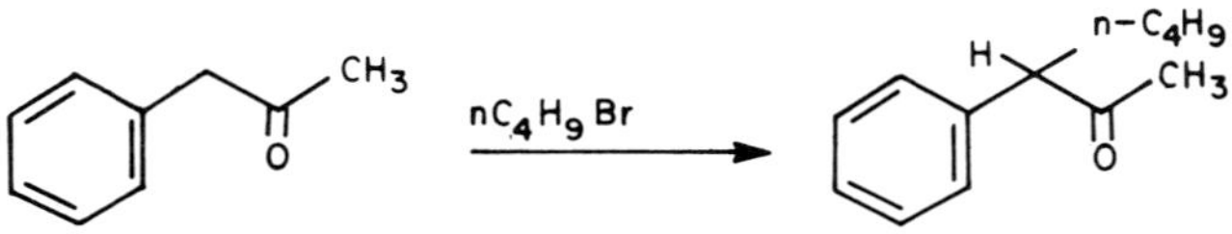

[62] + CH$_2$Br-C$_6$H$_5$ ⟶ [63]

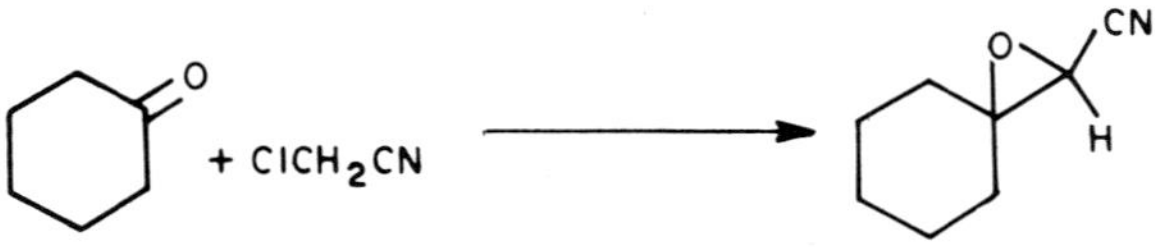

[64] + nC$_4$H$_9$Br ⟶ [65]

[66] + ClCH$_2$CN ⟶ [67]

alkylation of S-phenylglycolonitrile [54], [62] to [63], the alkylation of ketones [55], [64] to [65], and the Darzens condensation, [66] to [67] [56]. A similar technique, employing tetrabutylammonium hydroxide has been developed by Brandstrom and co-workers [57, 58, 59, 60, 61]. They have alkylated representative active methylene compounds such as acetylacetone, dimethyl benzoylmalonate and methyl acetoacetate in chloroform and methylene chloride organic phases with, among others, methyl, ethyl, and butyl iodides. In some cases [57, 58] of their method, called "extractive alkylation," the organic phase containing the tetrabutylammonium carbanion is separated from the aqueous phase but, like Makosza, they have found [59] that the alkylations may be just as readily conducted in the presence of water.

Starks has also investigated two-phase carbanion alkylation concentrating on displacement reactions of alkyl sulfonates and halides by cyanide ion in the presence of phosphonium salts.

The same theme runs through the procedures developed by these different sets of workers. A tetraalkylammonium chloride in catalytic quantity (1-5% of reactants) in strong aqueous base is mixed with an organic phase containing the base-sensitive substance (chloroform, malonic ester, etc.) and an alkylating agent, olefin, or other receptor of the base-generated reactant. The yields are high and secondary reactions are minimal. Many processes that cannot normally be carried out in homogeneous aqueous solution are readily accomplished in the two-phase system.

The exact mechanism of the phase transfer catalysis or extractive alkylation is not known. Evidence has been presented, however, by Herriot and Picker [62], that such reactions occur essentially in the organic phase and not at the interface or in micelles.

The work reviewed here concerns a very limited number of organic synthetic reactions. Organic chemists for the most part have not yet really applied phase-transfer catalysis as a general technique nor extensively investigated aqueous phase-transfer reactions to their fullest potential. It is clear, however, that these and other two-phase techniques offer simple and efficient means for conducting complex organic synthetic reactions.

REFERENCES

1. J. J. Bloomfield, *Tetrahedron Lett.*, 587, 591 (1968).
2. N. Kornblum and A. Lurie, *J. Amer. Chem. Soc., 81*:2705 (1959).
3. E. C. Taylor and A. McKillop, *Accts. Chem. Res., 3*:338 (1970).
4. R. L. Augustine, *Catalytic Hydrogenation*, Dekker, New York, 1965.
5. G. E. Ham, *J. Chem. Eng. Data, 8*:280 (1963).
6. H. O. House, *Modern Synthetic Reactions,* 2nd Ed., W. A. Benjamin, Menlo Park, 1972, Ch. 9.
7. F. M. Menger, *Chem. Soc. Rev., 1*:229 (1972).
8. R. P. Bell, *J. Phys. Chem., 32*:882 (1938).

9. G. A. Mansoori and A. J. Madden, *A.I.C.E. J.*, *15*:245 (1969).
10. F. M. Menger, *J. Amer. Chem. Soc.*, *92*:5965 (1970).
11. H. O. House, *Modern Synthetic Reactions*, 2nd Ed., Benjamin, Menlo Park, 1972, Ch. 5.
12. J. R. Conant and O. R. Quayle, *Organic Synthesis, Coll. Vol.*, *1*:211 (1941).
13. L. F. Fieser, *Organic Synthesis, Coll. Vol.*, *4*:197 (1963).
14. L. F. Fieser, *Organic Synthesis, Coll. Vol.*, *4*:189 (1963).
15. G. Rieveschl and F. E. Ray, *Organic Synthesis, Coll. Vol.*, *3*:420 (1955).
16. K. Bowden, I. M. Heilbron, E. R. H. Jones, and B. C. Weeden, *J. Chem. Soc.*, 39 (1946).
17. G. I. Poos, G. E. Arth, R. E. Beyler, and L. H. Sarett, *J. Amer. Chem. Soc.*, *75*:422 (1953).
18. R. H. Cornforth, J. W. Cornforth, and G. Popjak, *Tetrahedron*, *18*:1351 (1962).
19. J. C. Collins, W. W. Hess, and F. J. Frank, *Tetrahedron Lett.*, 3363 (1968).
20. J. Rocek and A. Riehl, *J. Amer. Chem. Soc.*, *88*:4749 (1966).
21. J. Rocek and A. Riehl, *J. Org. Chem.*, *32*:3569 (1967).
22. H. C. Brown, C. P. Garg, and K. T. Liu, *J. Org. Chem.*, *36*:387 (1971).
23. W. F. Bruce, *Organic Synthesis, Coll. Vol.*, *2*:139 (1943).
24. G. Rasmusson, H. O. House, E. F. Zaweski, and C. H. DePuy, *Organic Synthesis*, *42*:36 (1962).
25. E. W. Warnhoff, D. G. Martin, and W. S. Johnson, *Organic Synthesis, Coll. Vol.*, *4*:164 (Note 1)(1963).
26. W. S. Johnson, C. D. Gutsche, and D. K. Banerjee, *J. Amer. Chem. Soc.*, *73*:5464 (1951).
27. W. S. Johnson, in L. F. Fieser and M. Fieser, *Reagents for Organic Synthesis*, Wiley, New York, 1967, p. 143.
28. H. C. Brown and C. P. Garg, *J. Amer. Chem. Soc.*, *83*:2952 (1961).
29. F. H. Westheimer, *Chem. Rev.*, *45*:419 (1949).
30. An indication of the widespread utilization of this method may be obtained from citations of Ref. 28 in *Science Citation Index*.
31. H. C. Brown and A. Suzuki, *J. Amer. Chem. Soc.*, *89*:1933 (1967).
32. S. P. Acharya and H. C. Brown, *J. Org. Chem.*, *35*:3874 (1970).
33. D. J. Goldsmith and C. J. Cheer, *J. Org. Chem.*, *30*:2264 (1965).

34. B. E. Ratcliffe and C. H. Heathcock, *J. Org. Chem.*, *37*:531 (1972).
35. M. Barthelemy and Y. Bessiere-Chretien, *Tetrahedron Lett.*, 4265 (1970).
36. S. Naguchi, M. Imanishi, and K. Morita, *Chem. Pharm. Bull.* (Tokyo), *12*:1184 (1964).
37. A. E. Vanstone and J. S. Whitehurst, *J. Chem. Soc., C, 1966* (1972).
38. J. E. McMurry, *J. Amer. Chem. Soc.*, *90*:6821 (1968).
39. J. E. McMurry and S. J. Isser, *J. Amer. Chem. Soc.*, *94*:7132 (1972).
40. J. E. McMurry, *J. Org. Chem.*, *36*:2826 (1971).
41. C. M. Starks, *J. Amer. Chem. Soc.*, *93*:195 (1971).
42. W. Kirmse, *Carbene Chemistry*, 2nd Ed., Academic, New York, 1971.
43. M. Makosza and M. Wawrzyniewicz, *Tetrahedron Lett.*, 4659 (1969).
44. M. Makosza and E. Bialecka, *Tetrahedron Lett.*, 4517 (1971).
45. G. C. Joshi, N. Singh, and L. Pande, *Tetrahedron Lett.*, 1461 (1972).
46. I. Tabushi, Z. Yoshida, and N. Takahashi, *J. Amer. Chem. Soc.*, *92*:6670 (1970).
47. I. Tabushi, Z. Yoshida, and N. Takahashi, *J. Amer. Chem. Soc.*, *93*:1820 (1971).
48. M. Makosza, B. Serafin, and T. Urbanski, *Chem. Ind.* (Paris), *93*:537 (1965).
49. M. Makosza and B. Serafin, *Rocz. Chem.*, *39*:1223 (1965).
50. M. Makosza and B. Serafin, *Rocz. Chem.*, *39*:1401 (1965).
51. M. Makosza and B. Serafin, *Rocz. Chem.*, *39*:1799 (1965).
52. M. Makosza and B. Serafin, *Rocz. Chem.*, *39*:1805 (1965).
53. M. Makosza, *Tetrahedron Lett.*, 4621 (1966).
54. M. Makosza, E. Bialecka, and M. Ludwikow, *Tetrahedron Lett.*, 2391 (1972).
55. A. Jonczyk, B. Serafin, and M. Makosza, *Tetrahedron Lett.*, 1351 (1971).
56. A. Jonczyk, M. Fedorynski, and M. Makosza, *Tetrahedron Lett.*, 2395 (1972).
57. A. Brandstrom, P. Berntsson, S. Carlsson, A. Djurhuus, K. Gustavii, U. Junggren, B. Lamm, and B. Samuelsson, *Acta Chem. Scand.*, *23*:2202 (1969).

58. A. Brandstrom and U. Junggren, *Acta Chem. Scand.*, *23*:2203 (1969).
59. A. Brandstrom and U. Junggren, *Acta Chem. Scand.*, *23*:2204 (1969).
60. A. Brandstrom and U. Junggren, *Acta Chem. Scand.*, *23*:2536 (1969).
61. A. Brandstrom and U. Junggren, *Tetrahedron Lett.*, 473 (1972).
62. A. W. Herriot and D. Picker, *Tetrahedron Lett.*, 4521 (1972).

5.

KINETICS AND MECHANISMS

J. Howard Bradbury[*]	Peter J. Crawford[†]
Chemistry Department	*Commonwealth Scientific and*
Australian National University	*Industrial Research Organisation*
Canberra, A.C.T., Australia	*Campbell, A.C.T., Australia*

I.	INTRODUCTION	78
II.	TRANSPORT OF SOLUTE ACROSS LIQUID-LIQUID INTERFACES	79
	A. Absence of Stirring	79
	B. Presence of Stirring	80
III.	TRANSPORT PROBLEMS DURING INTERFACIAL SYNTHESIS	81
IV.	CHEMISTRY OF REACTIONS	83
	A. Mechanisms	83
	B. Solvent Effects	86
	C. Side Reactions	88
V.	KINETICS IN SPECIFIC SYSTEMS	90
	A. Preliminary Studies	90
	B. Reactions of Diacid Chlorides with Diamines	96
VI.	CONCLUSIONS	98
	REFERENCES	99

---

[*]Prepared during sabbatical leave at Laboratory of Molecular Biophysics, Department of Zoology, Oxford University, England.

[†]Present affiliation: Department of Environment, Housing, and Community Development, Canberra, A.C.T., Australia.

I. INTRODUCTION

An examination of the contents of this book, and of the earlier monograph by Morgan [1], shows that our expertise in the preparation of an amazing variety of polymers far exceeds our knowledge of the kinetics of interfacial polymerization processes. This is not surprising because of (1) the successful preparation by interfacial methods of many polymers using purely empirical reaction conditions, and (2) the complexity of the kinetics of interfacial reactions, which involve so many rate processes. Nevertheless, a deeper understanding of the kinetics and mechanisms of these processes is highly desirable, not only for its own sake, but also because it will undoubtedly lead to new developments in the field.

The kinetics of interfacial synthesis are intrinsically complex (as are all heterogeneous reaction systems) because of the large number of rate processes involved in the reaction. These include rates of mass transfer of reactants to the site of reaction and of products away from the site of reaction, as well as rates of chemical reaction in the region of an interface, in which surface effects (discussed by MacRitchie [2]) may be of considerable importance. The question of which process (if any) is much slower than the others, and hence can be considered to be rate-determining in any particular system, is very important and yet is difficult to resolve [3, 4].

In this chapter the processes of mass transfer in stirred and unstirred solutions are considered briefly (see also [5]), with particular reference to interfacial synthesis. This is followed by a discussion of the mechanisms of the chemical processes involved in certain interfacial reactions. Such mechanisms are often well understood because of the large amount of work done over many years on the reactions of analogous monofunctional reagents in homogeneous solutions. Consideration is then given to the few kinetic studies of interfacial polycondensation. In conclusion, the impression remains that a great deal more work is required in this challenging and interesting field.

5. Kinetics and Mechanisms

II. TRANSPORT OF SOLUTE ACROSS LIQUID-LIQUID INTERFACES

When a solute is transported across an interface from liquid A to liquid B the total resistance R to its passage is given simply as the sum of three terms R_A, R_B, and R_I, which represent, respectively, the resistances through the nonturbulent layers of liquids A and B adjacent to the interface I and through the monomolecular region constituting the interface [6]. These resistances are the reciprocals of the corresponding permeability coefficients k (cm/sec) which are defined by the equation for transfer across a plane surface, viz.

$$\frac{dq}{dt} = Ak \, \Delta c \tag{5.1}$$

where dq/dt is the rate of transfer in moles/sec of solute across a cross-sectional area A where there is a concentration difference Δc. The representation of the equation which is applicable to diffusional processes only (Fick's first law of diffusion) is

$$\frac{dq}{dt} = AD\frac{\Delta c}{\Delta x} \tag{5.2}$$

in which D is the diffusion coefficient of the solute and $\Delta c/\Delta x$ is the concentration gradient along which diffusion occurs. When transfer occurs by diffusion only it is seen from Eqs. (5.1) and (5.2) that

$$\frac{1}{R} = k = \frac{D}{\Delta x} \tag{5.3}$$

A. Absence of Stirring

In the simplest case in which a solute is diffusing from one liquid into another with both liquids taken as infinite in extent, the interfacial resistance term R_I (which is much smaller than the resistances for diffusion through liquids A and B) is neglected. It is possible to solve the second-order differential equations resulting from application of Fick's second law of diffusion and obtain a simple equation which relates the amount of material transported across

the interface with the change of concentration across it and the
diffusion coefficient of the solute.

However, even in this simple case there are complicating factors
that may affect the rate of transport, viz. (1) spontaneous emulsification at the interface [7], (2) interfacial turbulence [6], and (3)
interfacial convection [8]. The latter two effects result in increased rates of transport across unstirred interfaces by factors
ranging up to as high as 40-fold [8].

B. Presence of Stirring

The effect of stirring the liquid layers results in a large decrease
in the resistance terms R_A and R_B for the liquids. One approach is
to suppose that turbulent stirring keeps the composition of the
solute constant throughout the bulk of each phase up to a distance
Δx from the interface between the liquids and that laminar flow
occurs across the diffusion layer at the interface [9]. As the
degree of turbulence increases the thickness of the diffusion layer
decreases. In the limit of extreme turbulence the diffusional
process can no longer be considered as rate-determining so that
transfer would be governed by the equation

$$\frac{dq}{dt} = A \, \Delta C \, V_n \tag{5.4}$$

where V_n is the mean velocity of the liquid normal to the interface
and the other symbols have the same significance as in Eq. (5.1) [6].

It is noted that for diffusion-controlled transport, the rate
of transport is proportional to D (Eq. 5.2), whereas for extreme
turbulence (eddy diffusion) the rate of transport is independent of
D (Eq. 5.4). In practice, in stirred systems intermediate situations
occur in which the rate of transport is dependent on D^x, where x has
a value between 0 and 1 [6, 10]. Various empirical equations have
been developed that allow the calculation of the permeability or
transfer coefficient k (see Eq. 5.1) in different stirred cells
[6, 10, 11], and it is noted that k is decreased greatly at a surface covered with a protein monolayer [11, 12]. This is due to

5. Kinetics and Mechanisms

reduction in the turbulence near the interface, which results from the presence of the monolayer.

III. TRANSPORT PROBLEMS DURING INTERFACIAL SYNTHESIS

Any attempt to apply the principles developed in the preceding section to interfacial synthesis of polymers must take into account the overall characteristics of this process, and these will be discussed in general terms in this section.

The first question concerns the site of the chemical reaction. Experiments (1) with monofunctional compounds analogous to the difunctional monomers which produce polymers, (2) with mixed difunctional amines, (3) with observation of the effect of salts on rate constants, and (4) on direct observation of the polymer film in unstirred systems, indicate that the polymerization occurs in the organic phase adjacent to the interface [1, 3, 13, 14]. However there is an alternative proposal based on measurement of interfacial tensions, namely, that the polyamidation reaction occurs at the interface in a mixed monolayer containing the reactants [15]. In most systems examined up to the present time the weight of evidence certainly favors the former postulate. However it should be recalled that simple chemical reactions such as hydrolyses and esterifications in two-phase systems can take place either in monolayers at an interface [6] or homogenously in one of the phases [16], depending on the nature of the reactants, the solvents, concentrations, etc. It is therefore essential that in each interfacial polycondensation system studied kinetically a variety of techniques be used to establish the site of reaction and the role of the interface.

The second point is that the transport of reactants to the site of reaction can be considered, at least in principle, as a transport problem which involves an overall transfer coefficient k for each monomer (Eq. 5.1). However allowances must be made for the immobilization of some of the reactants as a result of chemical reaction, as the reaction proceeds. At the same time the transport of the products of reaction (which in polycondensation reactions are usually polymer molecules and small molecules) away from the site of reaction represents a closely related problem. Attempts have been made to describe systems involving diffusion with immobilizing chemical reaction as problems in diffusion. However, many conditions and simplifications have been imposed in these studies, and the treatments developed are only applicable to some selected systems [17-19].

A third question concerns the difficulty that arises if the polymer forms a coherent, insoluble film on the organic-phase side of the interface. Such a film will reduce greatly the rate of transport of reactants to the site of reaction. A related issue is the problem of dissolved polymer creating a substantial change in the viscosity and polarity of the solvent, thus influencing the diffusional processes and, at high concentrations in polar solvents, significantly affecting chemical rates.

Finally, if the chemical reaction is occurring at an interface between adsorbed monomers, then it becomes necessary to consider also the rate of adsorption of monomers at the interface and desorption of products [20], as well as the effect of the adsorbed material on the transfer coefficients of material through the interface.

It is clear that the rate constants for the chemical reactions will be greatly affected by the actual site of the reaction (whether in the organic phase or in the partially ordered region of the interface), by the presence or absence of a solid polymer film, and by any temperature effects due to heats of reaction. However discussion of the chemistry of the processes is reserved for subsequent sections. It is probably sufficient to note at this stage that the interfacial synthesis system is very complex and not completely understood and the application of the principles discussed in Sect. 5.II could only be achieved at a qualitative level.

IV. CHEMISTRY OF REACTIONS

A. Mechanisms

An examination of the kinetics and mechanisms of analogous homogeneous condensation reactions is of great importance to an understanding of the more complex interfacial polycondensation systems. It is not possible within the scope of this chapter to cover the basic chemistry of all relevant homogeneous condensation reactions, but by way of example the homogeneous reactions of acid chlorides with nucleophiles, such as amines, have been examined in some detail. This choice was influenced by the wealth of literature concerning amidation reactions, the number of studies of interfacial polyamidation reactions, and the commercial importance of polyamides. Broad coverage of other relevant condensation reactions such as esterification is given elsewhere [1, 21, 22]; and some of the most interesting recent mechanistic studies with one or more bifunctional intermediates have involved formation of ureas or urethanes [23-25].

It has been shown that reactions between acid chlorides and nucleophiles proceed essentially irreversibly [26] via a charged transition state [27-29]. The rate-determining step in such reactions can classically be regarded as either the bond-breaking of the carbon-halogen bond in the acid chloride (S_N1) or bond-making between the nucleophile and the carbonyl carbon atom (S_N2), depending on the experimental conditions of the reaction.

In highly polar solvents, in the absence of strong nucleophiles, the rate-determining step tends toward total bond-breaking as demonstrated in studies of the hydrolysis of p-substituted benzoyl chlorides in solvent mixtures in which the water content was progressively increased [30]. This work was extended to a limiting situation by Bender and Chen [31] who demonstrated that the neutral hydrolysis of mesitoyl chloride in acetonitrile-water (99:1) takes place via acylium ion formation.

On the other hand, in solvents of low polarity and in the presence of strong nucleophiles, bond-making has been shown to be rate-controlling [29] in confirmation of the results of an earlier extensive series of studies of bimolecular acylation reactions between aromatic acid chlorides and aromatic amines carried out by Hinshelwood and co-workers [28, 32-36]. These latter studies served to demonstrate that in a given solvent system, changes in rate constants of aromatic acylation reactions are determined almost exclusively by changes in activation energy, i.e., the entropy of activation is not significantly altered by the structure of the aromatic reactants.

Under suitable experimental conditions, solution condensation reactions of acid chlorides are known to be characterized by low activation energies [28, 34, 37]. Very low activation energies (1.3 kcal/mole) and high negative activation entropies (-57 cal/deg) have been observed in a study of the benzoylation of aniline in benzene-acetone mixtures [38].

Entelis and co-workers [39, 40] have used spectrophotometric means to follow the kinetics of hydrolysis and esterification of several (bifunctional) phthaloyl chlorides. In each case the reactions were shown to be first order in each reagent and to proceed via a transition state that was more polar than the reactants. It was also noted that both isoelectric activation energy [41] and second-order rate constants increased with increasing dielectric constant of the medium, indicating that entropy change was important in the formation of the transition state.

The reactions with amines of both mono- and bifunctional acid chlorides have been studied in n-heptane [42]. Rate constants were derived by spectrophotometric techniques. It was suggested that each reaction studied could be represented by the following simple sequence:

amine + acid chloride → ion-pair transition state → amide

It was claimed that the plots of observed rate of change of optical density vs. time were consistent with a rapid second-order reaction

5. Kinetics and Mechanisms

between the amino and acid chloride groups, followed by a slower first-order decay of the transition state to the product. However this claim was not adequately validated as was later recognised by the same research team [43]. In this connection it should be noted that: the experimental technique was not sufficiently sensitive to allow determination of the rate or order of the initial step or steps of the overall amidations; the simplifying assumptions made relating to reactivities of the various amino and acid chloride functional groups were not adequately supported theoretically or experimentally; and no account was taken of reports of the autocatalytic nature of amidation reactions in nonpolar solvents [32], nor of the nature of rate curves obtaining in such autocatalytic reactions.

A third set of experiments was performed by Entelis et al. [43, 44] using a stopped-flow technique to follow the kinetics. Reactions of mono- and bifunctional acid chlorides with mono- or bifunctional amines were studied in homogeneous and heterogeneous media. The kinetic curves obtained were of the same general shape for both homogeneous and heterogeneous reactions. The presence of several steps in each overall reaction was demonstrated. It was claimed that the initial portion of each curve was well described by a second-order equation. Rate constants calculated from these curves using initial reactant concentrations were of the order of 10^4 to 10^5 liter/(mole sec) at $20°C$. Although these rate constants are of the order of magnitude expected for such amidation reactions, as the authors themselves indicate, it would be necessary for further studies to be carried out before the steps detected in the above kinetic curves could be correlated with possible acylation schemes.

Of these approaches developed by Entelis et al. it appears that comprehensive studies by stopped-flow techniques of the kinetics of homogeneous amidation reactions involving one or more bifunctional reagents could provide useful basic data for future interfacial polyamidation studies. It seems less likely that such techniques could be adequately adapted to produce reliable data on heterogeneous systems.

B. Solvent Effects

1. pH and Ionic Strength
   of the Aqueous Phase

The chemistry and mechanisms of interfacial polycondensation reactions are profoundly affected by any variations in the composition of the aqueous phase during the reaction. For example, as the pH of the aqueous phase in an interfacial polycondensation reaction is altered, the availability of the diamine in reactive form changes. The extent of this change depends on the values of, and the changes in, the thermodynamic ionization constants of the diamine and the ionic strength of the solution. In fact, unless the pH and the ionic strength of the aqueous phase in an interfacial polycondensation reaction are maintained at constant values, changes in both these factors will affect the quantity of reactive diamine both in the aqueous phase and at the zone of reaction. Thus, the empirical optimum concentration ratios of reactants [1, 13] for formation of high polymer in numbers of interfacial systems are all dependent, among other factors, on the pH and ionic strength profiles of the aqueous phases throughout reaction.

The situation becomes even more complex in circumstances where an organic reactant or a charged polymeric monolayer may be adsorbed at a liquid-liquid interface during reaction, because the pH at the interface differs from that of the bulk aqueous phase. However, in common with the nonsurface active reactions described above, the pH at the interface and hence the course of the reaction would be influenced by variations in pH and ionic strength in the bulk phase [6]. It is interesting to observe that in some interfacial polycondensation systems the highest molecular weight polymer has been formed when the aqueous phase has been kept at an acid pH [45]. This result is chiefly attributable to use of diamines of low reactivity and to the reduced degree of hydrolysis of acid chlorides at low pH.

2. Salt Effects

Added salts are known to affect the activities of both initial and transition states in chemical reactions. The former interactions

5. Kinetics and Mechanisms

are usually described as thermodynamic salt effects and the latter as kinetic salt effects. Although theories relating to the thermodynamic effects of added salts on the activity coefficients of both electrolytes and nonelectrolytes present in aqueous solution are well developed [46], little opportunity exists to apply such sophisticated theories to the complex, nonideal and ever-changing environment presented by interfacial polycondensation systems. However two practical principles have been established concerning the extent to which reactants are salted in or out of aqueous phases by added salts: (1) Variations in the partition coefficients of diamines between organic solvents and water (containing added salts), at pH values where essentially only the free base exists, can be related to the empirically derived optimum concentration ratios for interfacial polycondensation systems, in which salts are present in the aqueous phases [1, 13]. (2) Acid chlorides are usually salted-out of the aqueous phase by small ions such as chloride ion [13, 47-49] but may be salted-in by large organic ions [47].

The effect of added salt on the rate of a chemical reaction has frequently been used as a means of identifying the underlying mechanism. In particular, it has been useful in determining the relative roles of bond-making and bond-breaking in the transition state of nucleophilic reactions [21]. However, in interfacial polycondensation systems where it is necessary to untangle the various effects of salts, either added prior to reaction or generated during reaction, it has been possible to use kinetic salt effects as mechanistic tools only in systems where physical variables have been carefully controlled [3].

3. Organic Solvents

The rates and mechanisms of condensation and polycondensation reactions in solution are greatly influenced by the choice of organic solvent—its polarity, dielectric constant, inertness to reactants and intermediates, and stability. Likewise the mechanism of reaction, molecular weight, and yield of polymer in interfacial polycondensations are very much governed by these factors as well as additional special factors such as the miscibility of the solvents with water, its possible role as an acid acceptor and its ability to dissolve the diamine initially contained in the aqueous phase.

Most of the systematic work on the influence of organic solvents on the products of interfacial polycondensation reactions has been performed by Morgan and co-workers [1]. They demonstrated that under comparable conditions to interfacial polyamidation the better the solvent for the polymer the higher the molecular weight of the product. To explain this result it has been postulated that in any given interfacial or solution polycondensation reaction a finite, reproducible solution period exists for the polymer in the organic solvent, during which time rapid chemical reaction in solution can continue until maximum molecular weight is attained and polymer precipitation ensues. This concept of a finite solution period is supported by studies [1, 50] in which metastable solutions of polyamides were prepared in chlorinated hydrocarbon solvents by both interfacial and solution polycondensation methods. The solution-precipitation technique described in Sect. 5.V.A represents a direct extension of this concept.

In practical terms most organic solvents that have limited solubility in water and are inert to reactive monomers have proved satisfactory for the preparation of high polymer by interfacial polycondensation methods. However, nonpolar solvents have been preferred for mechanistic studies in order to minimize both hydrolytic side-reactions and interference with spectrophotometric rate measurements by polymer dissolved in the organic phase [3, 51].

C. *Side Reactions*

There are several types of side-reactions commonly associated with polyamide formation by interfacial polycondensation, including hydrolysis and formation of cyclic or branched chain polyamides. Hydrolysis can have two basic effects on the polymeric product, depending on whether one or both functional groups of the initial reactant molecule are hydrolyzed. Thus, if one end-group is hydrolyzed, incorporation of the molecule in the forming oligomer chain

5. Kinetics and Mechanisms

leads to termination of polymer growth and to reduced polymeric molecular weight. If, on the other hand, both end-groups of the acid chloride are hydrolyzed, the concentration of monomer is effectively reduced, leading to a different ratio of reactant concentrations at the zone of reaction. Under most conditions used in interfacial polyamidation reactions, the former of these two alternatives would clearly prevail.

Hydrolysis usually takes place homogeneously in the aqueous phase during interfacial polycondensation [51, 52]. The relative rates of hydrolysis of a series of diacid chlorides and disulphonyl chlorides have also been examined in heterogeneous solvent systems [53-55]. Although some important variables were not maintained at constant values throughout reaction in these latter studies, it has been possible to demonstrate that the rates of hydrolysis were in general accord with the known chemical reactivities of the reactants examined. Data obtained from these studies of hydrolysis in two-phase systems have been used to explain decreases in molecular weight and yield of polymer with increasing temperature (or increasing hydroxyl ion concentrations) in the corresponding interfacial polymerization processes [53, 56].

Many cyclic products have been isolated from interfacial polycondensation reactions [1]. It is found that if one monomer is cyclic, the ring strain associated with formation of possible cyclic polymers reduces the tendency to form such cyclic products [57, 58]. Nonetheless a cyclic polyurethane has been prepared by interfacial synthesis involving the cyclic monomer piperazine [59].

Branched chain polyamides and polysulphonamides have been obtained [50, 60] from interfacial polycondensation reactions in which aliphatic primary amines were used. The branching results from acid chloride attack at the nitrogen of the amide or sulphonamide group initially formed in the polymerization process. It was found that the degree of branching increased with increasing concentration of acid chloride in the organic phase. The tendency for this sort of reaction to occur depends on factors such as the structure of the amine, the nucleophilicity of the nitrogen atom, the reactivity of the acid chloride, and the polarity of the solvent.

Under most conditions used in interfacial polyamidation reactions these latter two types of side-reaction lead to polymer of reduced molecular weight and to lower polymer yields. However, unlike hydrolysis, branching does not necessarily lead to chain termination and, accordingly, influences any empirically determined optimum concentration ratio of reactants in a given interfacial polyamidation system. It is conceivable under carefully controlled conditions that an examination of the extent of branching versus the extent of formation of linear polyamide could be a useful tool in a mechanistic examination; but, in general, it would seem more appropriate to select reactions that are known not to undergo side-reactions of this kind for any mechanistic study.

V. KINETICS IN SPECIFIC SYSTEMS

A. Preliminary Studies

1. Solution Polycondensation Reactions

Studies of essentially irreversible solution polycondensation reactions between glycols or diamines with diisocyanates, diacid chlorides or anhydrides have revealed a macroscopic order of reaction of approximately 2 and activation energies in the range 1-20 kcal/mole (see, e.g., Refs. 61-63). Methods of following the kinetics of reaction have included monitoring rate of release of gaseous byproducts such as hydrogen chloride [61] and colorimetric end-group determination at regular time intervals [63]. The studies have almost without exception been carried out at high reagent concentrations (0.1-5 M), and at temperatures well above room temperature. They have been characterised by divergence of rate constants from true second-order laws (on occasions throughout the reaction [64]). In a few cases, investigators [25, 65] have commented on the relationship between reactant activity and rates of reaction. Further, although it has been inferred in some of the studies that the results obtained, including

5. Kinetics and Mechanisms

the influence of organic solvents, are in accord with the known characteristics of bimolecular amidation reactions, the use of high concentrations and the lack of control of reaction parameters has rendered this group of experiments virtually meaningless as mechanistic studies.

There have also been many experiments in which the effects of variation of reactant concentrations and of ratios of reactants, solvents, and catalysts on polymer characteristics have been established. Perhaps the most interesting of these have been the so-called solution-precipitation polycondensation reactions studied by Turska et al. [66-68]. This method of polycondensation presents a possible way of producing polymer of desired molecular weight and distribution by the solution polycondensation technique. Thus if a solution polycondensation reaction at high temperature involving formation of a polycarbonate or polyarylate is allowed to proceed to an equilibrium state and then a quantity of nonsolvent for the polymer is added, the equilibrium is disturbed and precipitation of polymer commences. The precipitated polymer was shown also to be characterized by a narrow molecular weight distribution similar to a Gaussian distribution, whereas the polymer remaining in solution was characterized by a distribution similar to the most probable distribution. The minimum degree of polymerization at which polymer precipitated was determined by the experimental conditions prevailing in the particular system. Turska and co-workers were able to extend the work to show that, at constant time, after addition of a nonsolvent to a system in equilibrium, the higher the percentage of nonsolvent, the broader the molecular weight distribution and the lower the minimum degree of polymerization at which polymer precipitates. In each study excellent agreement was obtained between theoretical and experimental molecular weight distributions.

In most of the systems referred to above, as reaction proceeds many parameters are changing including site of reaction, solvent viscosity and polarity, polymer solubility, and monomer reactivities. Each change of this kind can influence the macroscopic kinetics of reaction significantly. Hamann et al. [69] have come to a comparable conclusion in their examinations of the kinetics of melt-polyesterification reactions. They point out that changes in polarity throughout reactions and lack of proportionality between reactant activities and concentrations at high monomer concentrations can lead, among other consequences to apparent fractional orders in a kinetic study instead of the correct integral ones. They overcame these problems in their studies by examining the kinetics toward the conclusion of the reactions when dilute solutions exist.

Thus the results of the studies described above serve only to provide a rough mathematical and/or empirical description of the particular system under examination. It is therefore not surprising to discover that in the numerous solution polycondensation studies performed over the past fifteen years, there has been little attempt at detailed kinetic or mechanistic analysis. An exception to this observation may be found in the mechanistic study by Lasocki and Dejak [70] of the solution polycondensations of silanediols. They investigated the acid- and base-catalyzed condensations in a series of careful experiments in various solvents. Although the underlying mechanism of the first step in these polycondensations was elucidated, the kinetics of the entire polycondensation process were not examined.

2. Interfacial Polycondensation Reactions

If we consider the example of interfacial polyamidation, the initial process consists of the reaction of diamine with diacid chloride to form oligomers. The rate of this reaction, and the subsequent reaction of oligomers to form polymer at the interface *may* be limited by the rate of transport of monomers to the interface (Sects. 5.II and 5.III), and in this respect the reaction differs greatly from the melt-polycondensation process in which equilibrium between the various species (monomer to polymer) is maintained throughout the reaction. The interfacial polyamidation may be terminated by (1) hydrolysis of an acid chloride functional group [51], (2) protonation of the amine group which would then be no longer accessible for further reaction,

5. Kinetics and Mechanisms

and (3) precipitation of the polymer, which will affect the rates of further reactions [71].

Fainberg and Mikhailov [72] attempted to follow the course of several polyamidation reactions in water-benzene systems by measurement of change of electrical conductivity of the aqueous phase. However under the experimental conditions selected the conductivity was influenced by formation of hydrochloric acid during reaction, disappearance of diamine, the dissolution of small quantities of water-soluble low-polymer and byproducts, and the presence of detergents, polymer precipitate, and droplets of organic phase in the aqueous phase. As a consequence such results as were obtained were uninterpretable. One of the major obstacles to this sort of approach to studies of interfacial systems lies in the complex interaction of electrical conductivity, aqueous phase polarity, pH, and diamine diffusivity.

Hodnett and Holmer [73] followed the rate of production of polymer in two stirred interfacial polyamidation reactions by weighing the polymer produced at varying time intervals. Each reaction was shown to be second order in nature when equivalents of reactants were initially present in the bulk phases. The rates of these reactions were much slower than those of a similar reaction studied by Katz [74] but this difference is readily attributable to differences in pH of the aqueous phases [1] and differing polymer solubilities. This approach was not able to yield basic information on the underlying mechanism of the interfacial polycondensation systems examined.

Sokolov et al. [4, 75-79] have carried out extensive studies of interfacial polycondensation reactions in gas-liquid [80] and liquid-liquid systems. In early studies they dealt with changes in polymer molecular weight and yield with a number of physical parameters, including interfacial tension and distribution coefficients of the monomers. In so doing they were able to demonstrate for gas-liquid systems that interfacial polycondensations occur in the immediate vicinity of the interface [76] and that for high molecular weight polymer relatively high gas-liquid surface tensions are necessary [75]. The work was extended to a theoretical treatment of copolycondensation in interfacial polycondensation systems [77, 79] involving reaction of two surface-active monomers A and B with a third reactant [80]. In this treatment they showed that the experimentally determinable copolycondensation constant (r) can be related theoretically to the initial and final concentrations of the monomers, $[A°]$, $[B°]$ and $[A]$, $[B]$, respectively, by the following formula:

$$\frac{[A]}{[A°]} = \left(\frac{[B]}{[B°]}\right)^r \tag{5.5}$$

They then derived relationships between this copolycondensation constant and other quantities such as Δn, the difference in the number of repetitive chain links between monomers A and B when the monomers belong to the same aliphatic homologous series.

All this work by Sokolov and co-workers has assisted in providing practical guidelines for synthesis of polymers, particularly by interfacial polycondensation at gas-liquid interfaces. The attempt to make the material even more useful by developing experimental and theoretical relationships involving the copolycondensation constant is to be commended. It must be noted, however, that the interfacial systems considered by Sokolov and co-workers have been far from ideal in mechanistic or thermodynamic terms. Consequently considerable further validation of assumptions, including those relating to ideal behavior of solutions at high concentration and temperature, would be necessary before the relationships derived could, as intended by these workers, be regarded as axioms for determining the controlling mechanism in a given interfacial copolycondensation reaction.

More recently Sokolov et al. [81] have indicated that some conclusions relating to mechanism of an individual polycondensation process can be made from an analysis of the dependence of the degree of polymerization on quantities such as the initial ratio of monomers, the relative rates of propagation to termination, and the concentration of monofunctional impurity. Although the simplifications in mechanism accepted in this study are too great, the suggestion that the dependence of the degree of polymerization on initial concentration of monofunctional, active additive will reveal details about the mechanism is worth pursuing.

5. Kinetics and Mechanisms

A second recent approach by Sokolov [4] to the assessment of mechanism in polycondensation processes is based on analysis of viscosity-polymer yield diagrams. It is suggested that a detailed examination of such diagrams for polycondensation systems can reveal the rate-controlling step in the system under examination. This study represents an extension of the earlier concept of optimum concentration ratios [1] for polycondensation reactions, but suffers from attempts to oversimplify the relationships between the many variables encountered in these systems. The statement made and applied in examining polycondensation systems, "that the basic laws governing irreversible polycondensation in homogeneous systems are analogous to those governing slow, reversible processes," is at variance with the work of Morgan [1] and of Bradbury et al. [3, 51, 71].

Carraher [82] found that the rate of formation of polyethylenesilylene by interfacial polycondensation followed a 1.67-order rate law. A model was proposed for the stirred system consisting of ethylene glycol droplets residing in the organic phase, and it was found that the results were consistent with an equation of the form

$$\text{Rate} = k[\text{spherical area of ethylene glycol}][\text{silane}]$$
$$= k[\text{ethylene glycol}]^{2/3}[\text{silane}] \quad (5.6)$$

However, this was a preliminary study involving a small number of experiments and use of relatively high reactant concentrations. Further experimentation would be necessary to demonstrate the validity of the assumptions that had to be made concerning surface and other physical variables in the system studied.

Finally, it should be noted that, in addition to the work of Sokolov and co-workers [77, 79, 80], other interfacial polyamidation reactions have been performed in which a second, competing diacid chloride or diamine has been added to the appropriate phase [83, 84]. In general, these studies have indicated that reactivities of diacid chloride or diamine play a secondary role to diamine diffusion. However, where the contribution of chemical reaction to the rate-controlling step has been significant, reaction-rate sequences have been shown to be in accord with known reactivities in homogeneous solution [1].

B. Reactions of Diacid Chlorides
   with Diamines

The studies discussed in Sect. 5.V.A have been considered as "preliminary" from the kinetic point of view, usually because of the lack of control (or constancy) of various parameters in the system during the course of the reaction. Some of the parameters which need to be controlled (or better still maintained constant) during a kinetic study are (1) the pH and ionic strength of the aqueous phase, (2) the temperature, (3) the degree of agitation of the heterogeneous system, (4) the occurrence of side-reactions such as the hydrolysis of acid chloride monomer, and (5) the formation of a film of polymer during the reaction.

It is possible to come to terms with these various factors as shown by a kinetic study of the reaction between stirred solutions of terephthaloyl chloride in n-heptane and piperazine in water [3]. The rate of the reaction is normally very fast, but it could be followed by normal procedures by use of low concentrations of reactants (10^{-2} to 10^{-6} M) and use of a low pH (about 6) which reduces very greatly the concentration of the active form (uncharged species) of piperazine. The low concentrations also minimized any kinetic effects due to the presence of insoluble polymer produced during the reaction. The much slower side-reaction (at pH 6) of hydrolysis of terephthaloyl chloride was monitored in a separate series of experiments [51], and a small correction was made for it. The rate of consumption of terephthaloyl chloride was followed spectrophotometrically.

The rate process for the consumption of terephthaloyl chloride is first order with respect to the acid chloride, but for piperazine the rate order decreases from 0.9 to 0.5 as its concentration increases. The rate of reaction increases with increasing pH (due to increase in the concentration of reactive uncharged piperazine molecules) and increasing temperature. The values of energies and

5. Kinetics and Mechanisms

entropies of activation are $\Delta E^{\ddagger}$ 4.1 kcal/mole and $\Delta S^{\ddagger}$ -59 cal/deg [3, 85] which are of comparable magnitude to those obtained in some homogeneous solutions for the benzoylation of aniline (Sect. 5.IV.A; Ref. 38).

Because of the relatively slow first-order rate constants for polyamidation (2-20 x 10^{-3} $\sec^{-1}$), it seems most unlikely that the rate of reaction is controlled by transport of reactants to the site of reaction, but rather by the chemical reaction itself. The formation of high polymer can only occur in a polycondensation if the rate of reaction between oligomers to form high polymer is at least comparable with the rate of reaction between monomers to produce dimer. This condition is satisfied in this polymerization because of the much lower pK of the amine-ended oligomers than of piperazine itself, which results in a higher level by a factor of about 100 of the concentration of the active (uncharged) species of the amine-ended oligomer as compared with the active (uncharged) piperazine.

Measurements of interfacial tensions showed the absence of specific adsorption of either monomer at the interface [51], and thus it was not necessary to consider the possibility of adsorption of monomers and desorption of products of reaction as possible rate-determining steps in the reaction (Sect. 5.III, Ref. 20). However in respect of other interfacial systems at high reactant concentrations appreciable adsorption of the aliphatic long chain diacid chloride sebacoyl chloride and aliphatic diamines (tetramethylenediamine, hexamethylenediamine, and decamethylenediamine) occurs at the water-organic phase interface [15, 86].

A simplified kinetic scheme was put forward in which the assumption was made that there are only two differently reactive types of acid chloride groups—viz., those attached to terephthaloyl chloride itself and those attached to polymeric species [3]. Similarly, the only two types of reactive amine groups are attached to the uncharged form of piperazine and to the uncharged form of an amine-ended polymeric species. On the basis of this scheme it is possible to explain the first-order dependence of the rate on terephthaloyl chloride concentration and the rate-order dependence of 0.5-0.9 on

amine concentration. However it was not possible to evaluate
explicitly any of the second-order rate constants given in the
kinetic scheme. This would only be possible if the dissociation
constants of piperazine and amine-ended oligomers in the buffer
solutions used were determined, and if techniques were devised to
measure the rate of consumption of both monomers during reaction.

VI. CONCLUSIONS

To summarize the present situation, it is clear that there have
been a number of kinetic studies which, with the exception of that
discussed in Sect. 5.V.B [3], have been of little value, because of
insufficient control of the variable parameters in the reaction.
Such control can and must be achieved before meaningful and reproducible kinetic results can be obtained.

However it is also obvious from Sects. 5.II and 5.III that
there are very difficult problems associated with the quantitative
interpretation of rates of transport of reactants and products of
reaction in these heterogeneous systems. It would therefore be
advisable in future kinetic studies to ensure that the rate of the
reaction is controlled by chemical reactions and not by transport
processes. This is not an easy matter to decide, but a chemically
controlled reaction rate should be sensitive to alterations of pH
of the aqueous phase and of other variables of this type, which
would be less likely to alter rates of transport. Rate studies in
such systems (hopefully in the absence of complications arising from
interfacial adsorption [2, 15]) should allow the evaluation of rate
constants for the polymerization process. These should be related
to the corresponding rate constants for reactions between monofunctional reagents in a homogeneous solution. As discussed in Sect.
5.IV, there is no doubt that much of the basic chemistry of interfacial polycondensation reactions can be elucidated by mechanistic
studies of model systems, involving reactions of monofunctional
reagents in single- and two-phase systems.

REFERENCES

1. P. W. Morgan, *Condensation Polymers by Interfacial and Solution Methods*, Interscience, New York, 1965.
2. F. MacRitchie, in *Interfacial Syntheses*, Dekker, New York, 1977, Chap. 6.
3. J. H. Bradbury, P. J. Crawford and A. N. Hambly, *Trans. Faraday Soc. 64*:1337 (1968).
4. L. B. Sokolov, *Polymer Science U.S.S.R., 12*:1097 (1970).
5. J. H. Rushton, in *Interfacial Syntheses*, Dekker, New York, 1977, Chaps. 2, 3.
6. J. T. Davies and E. K. Rideal, *Interfacial Phenomena*, Academic, New York, 1961.
7. J. T. Davies and J. B. Wiggill, *Proc. Roy. Soc., Ser. A, 255*: 277 (1960).
8. J. C. Berg and G. S. Haselberger, *Chem. Eng. Sci., 26*:481 (1971).
9. W. K. Lewis and W. G. Whitman, *Ind. Eng. Chem., 16*:1215 (1924).
10. G. R. A. Mayers, *Chem. Eng. Sci., 16*:69 (1961).
11. J. B. Lewis, *Chem. Eng. Sci., 3*:248, 260 (1954).
12. J. T. Davies and G. R. A. Mayers, *Chem. Eng. Sci., 16*:55 (1961).
13. P. W. Morgan and S. L. Kwolek, *J. Polym. Sci. 40*:299 (1959).
14. V. V. Korshak, T. M. Frunze, S. A. Pavlova, and V. V. Kurashev, *Vysokomol. Soedin., 5*:1130 (1963).
15. F. MacRitchie, *Trans. Faraday Soc., 65*:2503 (1969).
16. A. Viallard, *Chem. Eng. Sci., 14*:183 (1961).
17. J. Crank, *The Mathematics of Diffusion*, Oxford University, London, 1964.
18. P. V. Dankwerts, *Trans. Faraday Soc., 46*:300, 701 (1950).
19. R. A. Vroom and D. W. van Krevelen, in *Chemical Reaction Engineering, Proceedings of the Third European Symposium*, Pergamon, New York, 1965, p. 147.
20. F. M. Menger, *J. Amer. Chem. Soc., 92*:5965 (1970).
21. C. K. Ingold, *Structure and Mechanism in Organic Chemistry*, Cornell University, Ithaca, 1953.
22. J. S. Hine, *Physical Organic Chemistry*, 2nd Ed., McGraw-Hill, New York, 1962.
23. F. Willeboordse, *J. Phys. Chem., 74*:601 (1970).
24. S. Ito and Y. Tsutsumi, *Kogyo Kagaku Zasshi, 73*:1684 (1970).

25. O. G. Fedotova, A. G. Grozdov, and G. S. Kolesnikov, *Vysokomol. Soedin.*, *A9*, 1352 (1967).

26. S. G. Entelis, G. P. Kondrateva and N. M. Chirkov, *Vysokomol. Soedin.*, *3*, 1044 (1961).

27. J. J. Elliott and S. F. Mason, *Chem. Ind.*, 488 (1959).

28. H. S. Venkataraman and C. N. Hinshelwood, *J. Chem. Soc.*, 4986 (1960).

29. M. L. Bender and J. M. Jones, *J. Org. Chem.*, 7:3771 (1962).

30. D. A. Brown and R. F. Hudson, *J. Chem. Soc.*, 883 (1953).

31. M. L. Bender and M. C. Chen, *J. Amer. Chem. Soc.*, 85:30 (1963).

32. G. H. Grant and C. N. Hinshelwood, *J. Chem. Soc.*, 1351 (1933).

33. E. G. Williams and C. N. Hinshelwood, *J. Chem. Soc.*, 1079 (1934).

34. W. J. T. Pickles and C. N. Hinshelwood, *J. Chem. Soc.*, 1353 (1936).

35. A. N. Bose and C. N. Hinshelwood, *J. Chem. Soc.*, 4085 (1958).

36. H. S. Venkataraman and C. N. Hinshelwood, *J. Chem. Soc.*, 4977 (1960).

37. D. A. Brown and R. F. Hudson, *J. Chem. Soc.*, 3352 (1953).

38. E. Tommila and T. Vihavainen, *Acta. Chem. Scand.*, 22:3224 (1968).

39. S. G. Entelis, G. P. Kondrateva, and N. M. Chirkov, *Vysokomol. Soedin*, 3:1170 (1961).

40. S. G. Entelis, R. P. Tiger, E. Y. Nevel'skii, and I. V. Epel'baum, *Izv. Akad. Nauk SSSR, Otd. Khim. Nauk.*, 245:429 (1963).

41. E. S. Amis, *Kinetics of Chemical Change in Solution*, Macmillan, New York, 1949, p. 99.

42. S. G. Entelis and O. V. Nesterov, *Dokl. Akad. Nauk. SSSR*, *148*:1323 (1963).

43. S. G. Entelis, E. G. Bekhli, and O. V. Nesterov, *Kinetics and Catalysis, U.S.S.R.*, 6:281 (1965).

44. E. G. Bekhli, O. V. Nesterov, and S. G. Entelis, *J. Polym. Sci., Part C*, No. 16, 209 (1967).

45. L. V. Kozlov and T. V. Kudim, *Vysokomol. Soedin.*, 2:698 (1960).

46. R. A. Robinson and R. H. Stokes, *Electrolyte Solutions*, Academic, New York, 1955.

47. C. A. Bunton, N. A. Fuller, S. G. Perry, and I. H. Pitman, *J. Chem. Soc.*, 4478 (1962).

48. R. C. Kinstler, French Patent 1,211,287.

49. M. S. Akutin and L. A. Rodivilova, *Plast. Massy*, 14 (1960).

50. P. W. Morgan and S. L. Kwolek, *J. Polym. Sci.*, 62:33 (1962).

5. Kinetics and Mechanisms

51. P. J. Crawford and J. H. Bradbury, *Trans. Faraday Soc.*, *64*:185 (1968).
52. H. K. Hall, unpublished work, quoted in P. W. Morgan, *Condensation Polymers by Interfacial and Solution Methods*, Interscience, New York, 1965, p. 86.
53. W. J. Wasley, R. E. Whitfield, L. A. Miller, and R. G. Kodani, *Text. Res. J.*, *33*:1029 (1963).
54. V. V. Korshak, T. M. Frunze, S. V. Vinogradova, V. V. Kurashev, and A. S. Lebedeva, *Izv. Akad. Nauk SSSR, Otd. Khim. Nauk*, 1807 (1962).
55. R. G. Beaman, P. W. Morgan, C. R. Koller, E. L. Wittbecker, and E. E. Magat, *J. Polym. Sci.*, *40*:329 (1959).
56. T. M. Frunze, V. V. Korshak, V. V. Kurashev, and P. A. Alievskii, *Vysokomol. Soedin*, *1*:1795 (1959).
57. A. S. Shpital'nyi, Y. A. Kharit, R. B. Chernomordik, and D. G. Kulakova, *Zh. Prikl. Khim.*, *33*:1150 (1960).
58. A. S. Shpital'nyi, M. A. Shpital'nyi, D. G. Kulakova, Y. A. Kharit, and A. Y. Sorokin, *Zh. Prikl. Khim.*, *34*:408 (1961).
59. J. Farago, unpublished work, quoted in P. W. Morgan, *Condensation Polymers by Interfacial and Solution Methods*, Interscience, New York, 1965, p. 454.
60. S. A. Sundet, W. A. Murphey, and S. B. Speck, *J. Polym. Sci.*, *40*:389 (1959).
61. R. S. Velichkova, V. V. Korshak, S. V. Vinogradova, V. V. Ivonev, A. T. Ponomarenko, and N. S. Enikolopyan, *Izv. Akad. Nauk SSSR, Ser. Khim*, 858 (1969).
62. G. J. Howard and P. V. Wright, *J. Polym. Sci.*, Part A-1, *7*:2450 (1969).
63. E. Turska and A. Dems, *Polimery*, *15*:425 (1970).
64. S. V. Vinogradova, V. V. Korshak, G. S. Papava, N. A. Maisuradze, and R. S. Velichkova, *Izv. Akad. Nauk. SSSR, Ser. Khim*, *4*:820 (1970).
65. S. V. Vinogradova, V. A. Vasnev, V. V. Korshak, T. I. Mitaishvili, and A. V. Vasil'ev, *Dokl. Akad. Nauk SSSR*, *187*:1297 (1969).
66. V. V. Korshak, E. Turska, G. I. Timofeeva, and A. Dems, *Polimery*, *12*:169 (1967).
67. E. Turska and A. Dems, *J. Polym. Sci.*, Part C, No. 22, 407 (1968).
68. E. Turska and A. Dems, *Polimery*, *15*:432 (1970).
69. S. D. Hamann, D. H. Solomon, and J. D. Swift, *J. Macromol. Sci.-Chem.*, *A2*:153 (1968).

70. Z. Lazocki and B. Dejak, *Polimery*, *15*:391 (1970).
71. J. H. Bradbury and P. J. Crawford, *Proc. Roy. Aust. Chem. Inst.*, *36*:77 (1969).
72. E. Z. Fainberg and N. V. Mikhailov, *Vysokomol. Soedin.*, *2*:1039 (1960).
73. E. M. Hodnett and D. A. Holmer, *J. Polym. Sci.*, *58*:1415 (1962).
74. M. Katz, *J. Polym. Sci.*, *40*:337 (1959).
75. L. B. Sokolov and A. S. Astakhova, *Vysokomol. Soedin.*, *5*:176 (1963).
76. L. B. Sokolov and L. V. Turetski, *Vysokomol. Soedin.*, *6*:346 (1964).
77. L. B. Sokolov and L. V. Turetski, *Vysokomol. Soedin.*, *7*:1997 (1965).
78. L. B. Sokolov and T. J. Kudim, *Vysokomol. Soedin.*, *7*:1899 (1965).
79. L. B. Sokolov, V. Z. Nikonov, and G. N. Shilyakova, *Vysokomol. Soedin.*, *A11*:616 (1969).
80. L. B. Sokolov, in *Interfacial Syntheses*, Dekker, New York, 1977, Chap. 7.
81. L. B. Sokolov, *Vysokomol. Soedin.*, *Ser. A, 12*:901 (1970).
82. C. E. Carraher, *J. Polym. Sci.*, Part A-1, *7*:2359 (1969).
83. V. V. Korshak, T. M. Frunze, and L. V. Kozlov, *Izv. Akad. Nauk. SSSR, Otd. Khim. Nauk.*, 2062, 2226 (1962).
84. L. B. Sokolov and T. L. Kruglova, *Vysokomol. Soedin.*, *2*:704 (1960).
85. P. J. Crawford., Ph.D. Thesis, Australian National University, 1966.
86. V. Z. Nikonov and L. B. Sokolov, *Russ. J. Phys. Chem.*, *43*:581 (1969).

6.

INTERFACE EFFECTS ON CHEMICAL REACTION RATE

F. MacRitchie

C.S.I.R.O. Wheat Research Unit
North Ryde, N.S.W., Australia

I.	INTRODUCTION	104
II.	RATE OF ARRIVAL OF REACTANTS AT REACTION SITE	105
	A. Adsorption of Solute in Absence of Back-Diffusion	105
	B. Adsorption of Solute in Presence of Back-Diffusion	108
	C. Availability of Adsorption Sites	109
	D. Interfacial Pressure Barrier	110
	E. Electrical Potential Barrier	111
	F. Diffusion in Interface	112
III.	RATES OF REACTION IN INTERFACE	113
	A. Effects of Interface on Concentration and Orientation of Reactants	113
	B. Application of Theory of Absolute Reaction Rates to Reactions at Interfaces	119
	C. Effects of Interfacial Pressure and Electrical Potential	122
	D. Effects of Accessibility of Reactive Groups	123
	E. Action of Inhibitors	124
IV.	REMOVAL OF PRODUCTS FROM INTERFACE	124
	A. Diffusion in Absence of Barrier to Desorption	124
	B. Diffusion in Presence of Barrier to Desorption	125
	C. Interfacial Precipitation	126

V.	EXPERIMENTAL APPROACHES TO MEASUREMENTS OF RATES OF REACTION	127
VI.	EXPERIMENTAL RESULTS	130
	A. Effect of Interfacial Pressure	131
	B. Effect of Electrical Potential	132
	C. Effect of Accessibility of Reactants	133
	D. Reactions at Unidimensional and Zero-Dimensional Interfaces	135
	REFERENCES	136

I. INTRODUCTION

Phase boundaries occur abundantly in nature, and there is an equivalent chemistry for reactions at interfaces which is equally as fascinating as—and, in some respects, more so than—that for reactions in bulk. Although two-dimensional reactions have not been subjected to the same detailed study as those in three dimensions, their importance has been recognized in various branches of chemistry, including classical surface chemistry [1, 2], catalysis [3], and the chemistry of biological reactions [4]. The role that interfaces play in chemical reactions is becoming of increasing interest as methods are sought for the controlled synthesis of specific polymers.

Of course, a two-dimensional interface is not the only type of interface produced by phase boundaries. Unidimensional and zero-dimensional interfaces may and do occur. A simple example of a unidimensional interface is the boundary line formed in a beaker of liquid where the three phases, liquid, glass, and air, are in contact. If the beaker is made of vertical strips of different materials, zero-dimensional interfaces are formed at the junction of liquid, air, and the two types of solid. Unique energy fields are set up by all these types of interface, and this can lead to control of chemical reactions which is not possible in the three-dimensional state. In what follows, an attempt will be made to indicate how interfaces exert this control and what possibilities are thereby opened up. In general, the treatment will be directed toward processes occurring at two-dimensional

6. Interface Effects on Chemical Reaction Rate

interfaces, although in many cases the considerations will be applicable to other types of interface. Such interfaces need not be thought of as confined to boundaries between macroscopic phases. Macromolecules or micelles set up small-scale interfaces with their surrounding media, which may have great specificity as reaction sites for smaller molecules.

The overall rate of a reaction at an interface depends, firstly, on how quickly the reactants arrive at the reaction site; secondly, on the direct influence that the interface has on the reaction; and, thirdly, on the removal of products, thereby leaving the interface vacant for further reaction to occur. For any reaction, at a given time, the rate may be affected by all these three stages but will often be dominated by one of them. Each of the three stages will now be examined in detail.

II. RATE OF ARRIVAL OF REACTANTS AT REACTION SITE

A. Adsorption of Solute in Absence of Back-Diffusion

Molecules arrive at an interface from the bulk phase as a result of the random molecular movement known as diffusion. The average time (τ) that a molecule remains in its new environment depends on the magnitude of the free energy of adsorption, i.e., the difference in free energy of the molecule in its bulk and interface environments. The time τ may be estimated [5] from the equation:

$$\tau = \tau_0 e^{\frac{Q}{kT}} \tag{6.1}$$

where τ_0 is the time of oscillation of molecules in the adsorbed state, k is Boltzmann's constant, T is absolute temperature, and Q is the heat of adsorption. For molecules adsorbing from a gaseous phase on to solid surfaces, τ_0 is directly related to the time of vibration of the constituent atoms or molecules of the adsorbing surface and is of the same order of magnitude, namely 10^{-12} to 10^{-14} sec. For molecules adsorbing from liquid solution to liquid-gas and liquid-liquid interfaces, empirical results suggest a considerably greater value for τ_0. From a comparison of the adsorption energies (see Table 6.2) and the times of duration as spread monolayers for members of a homologous series such as the aliphatic alcohols, τ_0 would appear to have values in the order of 10^{-7} to 10^{-8} sec for this type of adsorption. This time range corresponds roughly to mean molecular displacements ($\bar{x}$), of 3-10 Å in the liquid phase, calculated by the Einstein diffusion equation, $\bar{x} = (2Dt)^{\frac{1}{2}}$, D being the diffusion coefficient and t the time.

If the time that a molecule is held in the interface is long compared to the time required for adsorption equilibrium to be reached, and provided that there is no barrier to the adsorption step, then, in the initial stages of adsorption, all molecules reaching the interface will be adsorbed and there will be no back-diffusion from the interface. Thus, almost immediately after formation of a new interface, the concentration of the subinterface layer (of several molecular diameters thickness) decreases to nearly zero owing to the migration of solute to the interface. A concentration gradient is set up in which molecules diffuse from the bulk region where they are at a constant concentration, C_0, to the sublayer which is effectively at zero concentration. Classical diffusion theory gives the number of molecules per unit area, n, diffusing to the sublayer in a time t [6] as:

$$n = 2C_0 \left(\frac{Dt}{\pi}\right)^{\frac{1}{2}} \tag{6.2}$$

where π is approximately 3.14. The rate of the diffusion process, which is therefore equal to the rate of adsorption dn/dt, follows by differentiation of Eq. (6.2):

$$\frac{dn}{dt} = C_0 \left(\frac{D}{\pi}\right)^{\frac{1}{2}} t^{-\frac{1}{2}} \tag{6.3}$$

Eq. (6.2) has been shown to describe very closely the initial stage of adsorption of proteins at a freshly formed interface [7]. Good agreement has been found for the reaction of ions (dissolved in the subphase) with stearic acid monolayers [8], for which Eq. (6.2) is also applicable.

6. Interface Effects on Chemical Reaction Rate

TABLE 6.1
Times Predicted by Eq. (6.2) for Close-packed Monolayers to form after Creation of a Fresh Interface

C_o (g/liter)	Octanol t calc.	BSA t calc.	BSA t meas.
3×10^{-2}	0.27 sec	7 sec	6 ± 1 sec
1×10^{-2}	2.5 sec	64 sec	55 ± 5 sec
3×10^{-3}	27 sec	11.9 min	9.0 ± 0.2 min
1×10^{-3}	4.2 min	107 min	42 ± 5 min

In Table 6.1 the times, calculated from Eq. (6.2), for the formation of close-packed monolayers, after creation of fresh interfaces, are tabulated for solutions of octanol and bovine serum albumin (BSA) at several values of the bulk concentration. The close-packed monolayers correspond to surface concentrations of 2×10^{14} molecules/cm^2 for octanol and 7.0×10^{-8} g/cm^2 for BSA. Values of 6.0×10^{-6} and 6.0×10^{-7} cm^2/sec are used in the calculations for the diffusion coefficients of octanol and BSA, respectively. For BSA the experimentally measured times are available [7] and are included in Table 6.1.

The validity of Eqs. (6.2) and (6.3) is dependent on the region through which diffusion occurs being free of convection. In general this is true; for it is found that, adjacent to interfaces, there is a layer of fluid in which laminar flow is observed and through which transfer of solute occurs as if the layer were stagnant. The thickness of this layer, usually referred to as the stationary layer or boundary layer for mass transfer, depends on the properties of the fluid and on conditions such as stirring or temperature control. Under certain conditions, the thickness of the stationary layer may be calculated using hydrodynamic boundary-layer theory [9, 10]. It may also be determined experimentally by measurements of rates of evaporation [11] or desorption [12]. At an air-water interface under normal conditions (i.e., no forced convection), the thicknesses of the stationary layers are of the order of 1 cm in air [13] and 1 mm in the water [12]. When the extent of the concentration gradient in an adsorption process becomes equal to the thickness of the stationary layer, the rate of adsorption thereafter becomes more rapid than predicted by Eqs. (6.2) and (6.3) owing to the influence of convection at the edge of the stationary layer. This effect is clearly shown in the results of Table 6.1 (cf. Refs. 7 and 8).

B. Adsorption of Solute in Presence of Back-Diffusion

Should the concentration of the sublayer become finite either due to desorption of adsorbed molecules because of a small value of τ or to failure of some of the molecules that reach the sublayer to adsorb, then a back-diffusion process from sublayer into the bulk occurs. Equation (6.2) then has to be extended to include an expression for back diffusion [14]:

$$n = 2(\frac{D}{\pi})^{\frac{1}{2}} \{C_o t^{\frac{1}{2}} - \int_o^{t^{\frac{1}{2}}} \phi(z) \, d[(t - z)^{\frac{1}{2}}]\} \qquad (6.4)$$

where z varies from 0 to t and $\phi(z)$ is the sublayer concentration at time z after surface formation.

The variation of $\phi(z)$ with time may be evaluated if it is assumed that equilibrium is maintained between interface and sublayer, i.e., that there is no activation energy barrier to adsorption. If this assumption is not valid, the presence of an energy barrier is immediately made evident by an impossibly high calculated value of D, using Eq. (6.4). The equilibrium interfacial tension-bulk concentration relations are measured for the solute and this data is used to convert the interfacial tension-time curve, measured under dynamic conditions, into a sublayer concentration-time curve. Ward and Tordai [14] have described a novel method which makes possible a direct graphical integration of the second term in Eq. (6.4).

Many cases of adsorption rates have been explained satisfactorily on the basis that diffusion is the only rate-determining process [15, 16]. In other cases, diffusion has accounted satisfactorily for the rate in the early stages of adsorption, but

6. Interface Effects on Chemical Reaction Rate

activation energy barriers have been evident in the later stages [15, 17].

Three main types of free energy barrier to adsorption have been clearly elucidated. These are concerned with the unavailability of sites in the case of localized adsorption, the work of overcoming an interfacial pressure barrier in mobile adsorption, and the work of overcoming an electrical potential barrier, the latter operating for charged molecules in both localized and mobile adsorption (see Fig. 6.1).

C. Availability of Adsorption Sites

At solid-fluid interfaces where adsorption may be localized at fixed sites, molecules reaching the interface may find sites at the point of contact to be occupied so that these molecules will be returned to the bulk or may remain in the sublayer until an unoccupied site is reached. If, at a given time, the fraction of sites occupied is θ, the rate of adsorption will be given by [18]:

$$\frac{dn}{dt} = k_1 C_o (1 - \theta) \tag{6.5}$$

where k_1 is the rate constant for the adsorption.

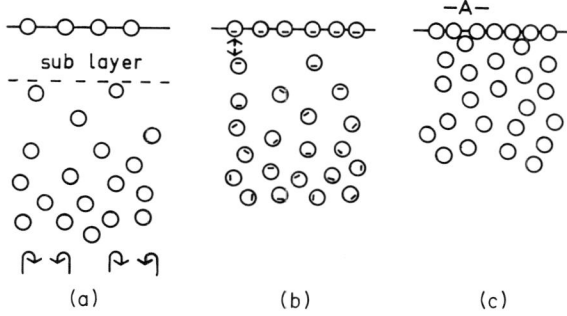

FIG. 6.1. Successive stages in an adsorption process. (a) Diffusion governed by Eq. (6.2); (b) repulsion of charged molecules as a result of electrical potential at interface (c) molecules need to compress monolayer against the interfacial pressure in order to create a space of area A to enter.

D. *Interfacial Pressure Barrier*

At many interfaces with solids and certainly at all gas-liquid and liquid-liquid interfaces, there are no fixed adsorption sites. However, adsorption of molecules at these interfaces produces an interfacial pressure which gives rise to an energy barrier. The interfacial pressure (Π) is defined as the difference between the interfacial tensions of the clean (γ°) and monolayer covered (γ) interfaces:

$$\Pi = \gamma^\circ - \gamma \tag{6.6}$$

Π has units of a two-dimensional pressure (dynes/cm) and is completely analogous to three-dimensional pressure.

In order for a molecule to enter the interfacial film from the bulk phase, it must do work of compression against the molecules already adsorbed so as to create an area of interface (A) for it to move into [17]. This amount of work is given by $\int_0^A \Pi \, dA$. If the adsorption step is rapid enough so that it may be considered to occur effectively at constant Π, then the work term is simply ΠA. The rate of adsorption (dn/dt) will be equal to the number of molecules striking unit area of interface in unit time ($k_1 C_0$) and having energies in excess of ΠA:

$$\frac{dn}{dt} = k_1 C_0 e^{-\Pi A/kT} \tag{6.7}$$

The area A has been shown to agree closely with the area occupied by the molecule in a close-packed monolayer [19]. For polymers, A may correspond to the area of a segment rather than the entire molecule [19]. This will occur with flexible polymers where the polymer chain behaves essentially as a number of independent kinetic units [20, 21]. Simple detergent-like compounds such as fatty acids and alcohols have A values of about 20 Å^2 at an air-water interface. For these compounds, ΠA has a value roughly equal to kT at an interfacial pressure of 20 dynes/cm, so that the interfacial pressure barrier is of comparatively minor importance. On the other hand, for larger molecules or polymers, A may have values

6. Interface Effects on Chemical Reaction Rate

of 100 $\overset{\circ}{A}{}^2$ and greater [19]. For A = 100 $\overset{\circ}{A}{}^2$, ΠA would have a value of about 5 kT at 20 dynes/cm, which would mean that only 1 in 100 of the molecules striking the surface would have sufficient energy to adsorb. It is readily seen that, for competition between small and large molecules, adsorption of the smaller molecules becomes relatively more favorable as the interfacial pressure increases. Similar considerations might explain why Defay and Hommelen [15] found that adsorption of monocarboxylic acids could be explained satisfactorily by diffusion theory, whereas an activation energy barrier needed to be invoked for adsorption of a dicarboxylic acid in the later stages.

E. Electrical Potential Barrier

The electrical potential barrier is the most general energy barrier to adsorption in the sense that it may act in cases of both localized or mobile adsorption. On the other hand, it only assumes importance where the adsorbing molecules carry an electrical charge.

The adsorption of charged molecules at an interface gives rise to a concentration of charges in the plane of these groups. This produces a high electrical potential (ψ) relative to any point in the adjacent bulk phase, although the effect is counteracted to some extent by the tendency of oppositely charged ions and molecules to congregate around the charged plane. In this way an electrical double layer is set up, consisting of a fixed layer of adsorbed charges and an adjacent diffuse layer of charges predominantly of opposite sign. For a molecule of charge q to diffuse from the bulk into the interface, a quantity of work equal to $\int_0^\psi q\, d\psi$ must be done. The rate of adsorption, in the absence of other barriers, will be given by:

$$\frac{dn}{dt} = k_1 C_o e^{-q\psi/kT} \tag{6.8}$$

An alternative and equally valid approach is to consider that the concentration in the sublayer is not C_o, the concentration in bulk, but $C_o e^{-q\psi/kT}$. In general, the adsorbing molecules will have the same sign of charge as ψ since the potential is produced by these same charged molecules. However, it may happen, such as in cases of adsorption into mixed monolayers, that the charge on the adsorbing molecule is opposite in sign to ψ. For this case, the rate of adsorption will be accelerated and the effective sublayer concentration will be given by $C_o e^{+q\psi/kT}$.

Much theoretical work has been concerned with describing the electrical conditions at interfaces, including evaluation of the potential at different points near the interface [22, 23]. These potentials are not amenable to direct measurement, although Davies has made progress in their evaluation using experimental methods [24]. Hartley and Roe [25] have suggested that the potential ψ to be used in Eq. (6.8) closely approximates to the ζ potential which is measured in electrokinetic experiments. Some confirmation of Eq. (6.8) and of the suggestion of Hartley and Roe was obtained from a study of rates of adsorption of the protein lysozyme at charged interfaces [26]. The logarithm of the rate varied linearly with the ζ potential, measured independently, while the slope of the line gave a reasonable value for the electrophoretic charge of the protein. Values of 50 mV for ψ are common at charged interfaces, but this may rise to 250 mV in certain special cases. The magnitude of the resulting energy barriers would be about 2kT (50 mV) and 10kT (250 mV) for an adsorbing molecule carrying one electronic charge having the same sign as ψ. The proportion of successful collisions leading to adsorption would be roughly 1 in 7 and 1 in 2×10^4, respectively.

F. Diffusion in Interface

Reactions at interfaces may occur between adsorbed molecules and species in solution. At solid interfaces, reaction may occur between adsorbed species and the adsorbent as in chemisorption. Alternatively, reaction may occur between molecules of the same or different species adsorbed at the interface. In the latter case, in order to react, molecules (after adsorption) must come together by an interfacial diffusion process.

6. Interface Effects on Chemical Reaction Rate

Where adsorption is said to be localized at sites, this may not necessarily mean that no migration of molecules in the interface occurs. The term localized adsorption means only that the interfacial mobility is small compared to the rates of adsorption or desorption, so that species arriving at or leaving the interface do so relative to an essentially unchanging surface environment. If the activation energy for migration from one site to another is appreciable but less than that for desorption, the adsorption is still localized but the layer is mobile. It is because of this mobility that bimolecular interfacial reactions can occur in these cases. For adsorbed films of mobile molecules such as at gas-liquid and liquid-liquid interfaces, the same laws of diffusion apply as for three dimensions, modified only by the loss of the one degree of translational freedom. The diffusion coefficient at a gas-liquid interface might be expected to be intermediate between those in the bulk phases. Few attempts have been made to measure diffusion coefficients in the interface [27], although the problem has been considered in studies of interfacial reactions [28]. Diffusion at interfaces may be a limiting factor for reaction rates in cases where the interfacial viscosity reaches very high values such as in polymer films.

III. RATES OF REACTION IN INTERFACE

A. Effects of Interface on Concentration and Orientation of Reactants

The rate of a chemical reaction depends on the concentrations of reactants and the specific rate constant for the reaction, both of which may be greatly altered when reaction occurs at an interface instead of in the bulk phase. Near any phase boundary, it is necessary to consider the molecular distribution of species between three phases—the two adjacent bulk phases and the interfacial phase. At a hydrocarbon liquid-water interface, for example, molecules of n-octane will distribute themselves preferentially in the hydrocarbon phase whereas hydroxyl ions will concentrate almost exclusively in the aqueous phase. This is because a molecule of n-octane has a much lower free energy in a hydrocarbon than in water, and vice versa for the hydroxyl ion. However, a molecule of n-octanol, which is made up of an n-octane chain and a hydroxyl group, will be more concentrated at the interface than in either of the two bulk phases; for here the octane chain can interact with the hydrocarbon liquid and the hydroxyl group with the water, the configuration of lowest free energy in the system. The interface therefore acts as a unique type of solvent, a specific two-dimensional solvent for this type of compound.

The molecular distribution between bulk and interface (n_1/n_2) is given by the equation

$$\frac{n_1}{n_2} = \frac{p_1}{p_2} e^{\lambda/kT} \tag{6.9}$$

where p_1 and p_2 are the a priori probabilities of finding a molecule in bulk and interface, respectively, and λ is the energy difference at absolute zero between molecules in bulk and at the interface. If the adsorbed monolayer is subjected to an interfacial pressure (Π) or an electrical potential different to that in bulk (ψ), additional terms must be included:

$$\frac{n_1}{n_2} = \frac{p_1}{p_2} e^{\lambda/kT} e^{\Pi A/kT} e^{q\psi/kT} \tag{6.10}$$

To illustrate the application of Eq. (6.10), the distribution of some normal aliphatic alcohols between aqueous solution and the air-water interface have been calculated. In order to evaluate λ, use was made of the principle of independent surface action, originally proposed by Langmuir [29]. This supposes that the energy per unit area of a group in an isolated molecule immersed in a solvent is the same as the interfacial energy between a macroscopic phase of the same groups and the solvent. Interfacial energies are obtained by extrapolating the interfacial tension-temperature relationship to the temperature of absolute zero. This gives values of 117 ergs/cm^2 for the air-water interface ($\gamma^°_{AW}$), 65 ergs/cm^2 for a hydrocarbon-

6. Interface Effects on Chemical Reaction Rate

water interface (γ_{HW}^o), and 51 ergs/cm^2 for a hydrocarbon-air interface (γ_{HA}^o). Because γ_{HA}^o is lower than γ_{HW}^o, an alcohol molecule can lower its energy by adsorption at an air-water interface, thus replacing some of its hydrocarbon-water interface by hydrocarbon-air. In the process, an area of air-water interface is eliminated, giving a further significant lowering of energy to the system.

To obtain λ, volumes of molecules were calculated and the radii (r) derived, assuming the molecules to be close-packed spheres. In the adsorbed state, molecules were assumed to be half immersed in the water, this half of the molecule containing the OH group. λ was then equal to $-[½ \cdot 4\pi r^2 (65 - 51) + \pi r^2 \cdot 117]$ ergs. For mobile adsorbed molecules, p_2 will differ from p_1 by virtue of lacking a partition function corresponding to one degree of translational motion, $(2\pi mkT)^{½}/h$, which has values close to 10^9 for most molecules at room temperature. It will also lack a partition function for one degree of rotation about an axis in the plane of the interface. This is given by $(8\pi^2 IkT)^{½}/h$, which usually has values between 10 and 100. Table 6.2 summarizes the calculated values of p_1/p_2 and $e^{\lambda/kT}$ for the distribution of several alcohols of different chain lengths between water and the air-water interface.

The distributions between bulk and interface may now be calculated using Eq. (6.10) and compared with the experimental data of Posner et al. [30]. Several values are compared in Table 6.2. For calculating n_2, the appropriate value of Π was taken from the experimental data [30] and a constant value of 20 Å^2 used for A in the term $e^{\Pi A/kT}$. The agreement between calculated and experimental values of n_2 is remarkably close in view of the simplified treatment and tends to show that many of the assumptions made are very good approximations. However, the molecular shapes and configurations which have been assumed are only applicable to compounds giving expanded films and need to be modified for gaseous or condensed monolayers. In addition, other factors need to be taken into account in a rigorous treatment of the distribution between bulk and interface. The aim here was to illustrate the main principles involved,

TABLE 6.2

Relevant Data for the Distribution of
Aliphatic Alcohols between Bulk and Interface at 25°C

Alcohol	Volume (Å^3)	$\dfrac{p_1}{p_2}$	$e^{-\lambda/kT}$	n_1 (molecules/cm^3)	n_2 calc. (molecules/cm^2)	n_2 expmtl. (molecules/cm^2)
Methanol	67	1.78×10^{10}	1.00×10^3			
Butanol	153	5.46×10^{10}	1.95×10^5	1.0×10^{19}	3.1×10^{13}	7.2×10^{13}
				1.6×10^{20}	1.6×10^{14}	3.2×10^{14}
Octanol	263	1.13×10^{11}	3.39×10^7	1.1×10^{17}	2.7×10^{13}	9.6×10^{13}
Tetradecanol	427	2.18×10^{11}	1.59×10^{10}	1.8×10^{18}	0.8×10^{14}	4.0×10^{14}

6. *Interface Effects on Chemical Reaction Rate*   117

and Langmuir's approach is very useful for this purpose. More sophisticated treatments are also available for evaluating λ [31, 32].

If we assign a volume to the interfacial phase by assuming a thickness approximately equal to the thickness of the adsorbed monolayer, it is evident that even for butanol (10 Å thickness), the concentration in the interface is greater by a factor of 10 than the equilibrium concentration in bulk. For octanol, the ratio of interfacial to bulk concentration at equilibrium is of the order of 10^3, and as λ increases the distribution ratio becomes more spectacular.

Besides the property of concentrating reactants, another general effect of an interface, which influences reaction rates, is that of orienting molecules. For certain reactions, molecules must come together with a specific orientation in order to react. In bulk solution, collisions between molecules occur with random orientation. At an interface, the relative probability of a given orientation (P_a/P_b) may be calculated by a form of Eq. (6.9) with (n_1/n_2) replaced by P_a/P_b. Considering the two possible orientations of the methanol molecule shown in Fig. 6.2, the a priori probabilities of either

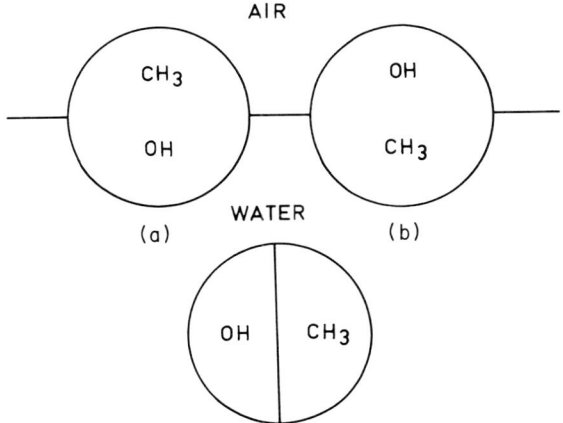

FIG. 6.2. Different orientations of a methanol molecule at an air-water interface.

orientation are equal. Taking values of 190 ergs/cm^2 and zero for the energies, at absolute zero, of hydroxyl groups with air and water, respectively, the value obtained for P_a/P_b is 3.5 x 10^8. The orientation of molecules at interfaces is confirmed by many results; for example, the surface tensions of alcohols are practically identical to those of paraffin hydrocarbons.

Increased reaction rates as a result of greater reactant concentrations and specific orientation of reactive groups may have far-reaching effects on reaction mechanisms. For example, if the rate of the chain-propagating step in a polymerization is greatly increased while that of chain-termination remains the same, much higher molecular weight products should be obtained. In addition, the configurational constraint imparted to an adsorbed molecule may profoundly affect the nature of a reaction. Carothers [33] in 1936 drew attention to the fact that polymerization of α-amino acids, in vitro, produced cyclic compounds whereas, in living organisms, reaction is exclusively intermolecular and linear polymers result. If reaction, in the latter case, were preceded by adsorption, thus fastening head, tail, and middle of the molecule to the interface, intramolecular approaches might then not be possible. It is noteworthy, in this connection, that one-dimensional interfaces would preclude, with complete certainty, any reactions other than between terminal groups of different molecules. Another constraint that may be imposed by an interface is that its area be limited to microscopic dimensions such as in the case of an emulsion droplet. This, again, could be important for polymerization reactions, for the chain-terminating step might be equivalent to adsorption saturation. One way of producing a polymer with a narrow molecular weight distribution could then be to polymerize on a homogeneous array of such interfaces.

6. *Interface Effects on Chemical Reaction Rate* 119

B. *Application of Theory of Absolute Reaction Rates to Reactions at Interfaces*

The theory of absolute reaction rates [34] seeks to explain quantitatively the rates of reactions on the basis of fundamental molecular properties. The theory postulates that for reaction to occur, reactants must form a critical configuration (transition complex) which then decomposes spontaneously to give the reaction products. An equilibrium is assumed between reactants and transition complexes, the reaction rate being determined by the concentration of transition complexes and the rate at which they pass through the critical configuration. The general expression for the rate of any chemical reaction or rate process (omitting the transmission coefficient) is then given, in terms of the concentrations of reactants ($C_A, C_B, \ldots$), the free energy of activation ($\Delta F^\ddagger$), and a universal frequency (kT/h) which includes Boltzmann's constant and Planck's constant, as:

$$\text{Rate} = (C_A C_B \ldots) \frac{kT}{h} e^{-\Delta F^\ddagger/RT} \tag{6.11}$$

which may be expanded as follows:

$$\text{Rate} = (C_A C_B \ldots) \frac{kT}{h} \frac{F^\ddagger}{F_A F_B} e^{-E_0/RT} \tag{6.12}$$

This expression includes the partition functions per unit volume for the transition complex ($F^\ddagger$) and reactants ($F_A, F_B, \ldots$) and the activation energy of the reaction at absolute zero (E_0) i.e., the difference at absolute zero between the energy per mole of activated complex and the sum of the energies of the reactants.

For the purpose of the present discussion, it will be assumed that the activation energy is the same for the bulk reaction as for the reaction at the interface. The theory of absolute reaction rates will now be used to compare the rate of a bimolecular reaction in bulk solution with the rates of the same reaction at different types of interface. The reaction to be considered is of the type:

$$A + B \rightleftharpoons AB^\ddagger \rightarrow C \tag{6.13}$$

1. Reactions Between Mobile Adsorbed Reactant Molecules at Two-Dimensional Interfaces

The solution contains 10^{18} molecules/cm^3 of both A and B. The concentrations at the interface are also equal (10^{14} molecules/cm^2). It will be assumed that the partition functions for one degree of translation (f_t) have values of 10^9 and the partition functions for rotation (f_r) have values of 10. The reactants and transition complex in the interface reaction have one degree of translational freedom and two degrees of rotational freedom less than in bulk solution. The rates at the interface (rate$_i$) and in bulk (rate$_b$) are

$$\text{Rate}_i = C_A C_B \frac{kT}{h} \frac{f_t^2 f_r}{f_t^2 f_r f_t^2 f_r}$$

$$= \frac{kt}{h} \cdot 10^{14} \cdot 10^{14} \cdot \frac{1}{10^{19}} \quad (6.14a)$$

$$\text{Rate}_b = C_A C_B \frac{kT}{h} \frac{f_t^3 f_r^3}{f_t^3 f_r^3 f_t^3 f_r^3}$$

$$= \frac{kT}{h} \cdot 10^{18} \cdot 10^{18} \cdot \frac{1}{10^{30}} \quad (6.14b)$$

We now compare the absolute rates for two systems that contain the same numbers of molecules. There will be 10^{14} molecules in a portion of the bulk solution 1 cm^2 × 10^{-4} cm. Therefore,

$$\frac{\text{Rate}_i}{\text{Rate}_b} = \frac{10^{14} \cdot 10^{14}}{10^{19}} \cdot \frac{10^{30}}{10^{-4} \cdot 10^{18} \cdot 10^{18}}$$

$$= 10^7 \quad (6.15)$$

2. Reactions Between Adsorbed Mobile Molecules at a One-Dimensional Interface

Reactant concentrations are 10^{18} molecules/cm^3 in bulk and 10^7 molecules/cm at the interface. Molecules at the interface retain partition functions for single degrees of translation and rotation.

6. Interface Effects on Chemical Reaction Rate

$$\text{Rate}_i = C_A C_B \frac{kT}{h} \frac{f_t^\ddagger}{f_t f_t}$$

$$= \frac{kT}{h} \cdot 10^7 \cdot 10^7 \cdot \frac{1}{10^{10}} \tag{6.16}$$

The number of molecules in 1 cm of interface (10^7) would occupy a volume in bulk of 1 cm^2 × 10^{-11} cm. Therefore,

$$\frac{\text{Rate}_i}{\text{Rate}_b} = \frac{10^7 \cdot 10^7}{10^{10}} \cdot \frac{10^{30}}{10^{-11} \cdot 10^{18} \cdot 10^{18}}$$

$$= 10^9 \tag{6.17}$$

3. Reactions Between Molecules Adsorbed on Fixed Sites at a Two-Dimensional Interface

In this case, reactants and transition complex at the interface lose all degrees of translational freedom and possibly all of rotation:

$$\text{Rate}_i = C_A C_B \frac{kT}{h} \tag{6.18a}$$

$$\frac{\text{Rate}_i}{\text{Rate}_b} = 10^{26} \tag{6.18b}$$

4. Effect of Change in Activation Energy

If the activation energy is reduced by 6 kcal/mole at room temperature, there will be an increase in reaction rate by a factor of about 10^4. The previous three examples illustrate the effect that an interface is likely to have on a reaction by altering the reactant concentrations and the entropy of activation. In systems where there are large interfacial areas concentrated in small volumes, such as in living cells, these factors will be of great importance. On the other hand, where the extension of interface is small compared to the volume of bulk phases, the increase in rate caused by an interface may be more than compensated by the very much greater amounts of reactants present in the bulk phases. Only by significantly reducing the activation energy could the interface reaction then assume any importance.

It has been concluded from reactions studied to date [3] that experimental activation energies are not significantly different between bulk and interface. However, reliable values for activation energies of interface reactions are difficult to measure, especially very fast reactions, so the question of whether differences do exist remains open.

C. Effects of Interfacial Pressure and Electrical Potential

To allow for the influence of external parameters on reaction velocity, Eq. (6.11) can be written in the form

$$\ln k = \text{const.} - \frac{1}{RT} \left[\frac{\partial (\Delta F^{\ddagger})}{\partial \chi}\right]_T d\chi \tag{6.19}$$

where χ is some continuous parameter representing an intensity factor such as interfacial pressure or electrical potential, the assumption being made that $\partial (\Delta F^{\ddagger})/\partial \chi$ remains constant over a range of this parameter, and k is the rate constant for the reaction.

In the case of interfacial pressure (Π),

$$\frac{\partial (\Delta F^{\ddagger})}{\partial \Pi} = \Delta A^{\ddagger} \tag{6.20}$$

where $\Delta A^{\ddagger}$ is the difference in area between the activated complex and the reactant molecules. It follows that:

$$\frac{d(\ln k)}{d\Pi} = -\frac{\Delta A^{\ddagger}}{RT} \tag{6.21}$$

It can be seen that the effect of interfacial pressure on reaction velocity depends on the sign as well as the magnitude of $\Delta A^{\ddagger}$. If $\Delta A^{\ddagger}$ is negative (i.e., there is a decrease in area when the activated complex forms from the reactant molecules), increase of interfacial pressure should increase the rate, and conversely for a positive $\Delta A^{\ddagger}$.

Where the intensity factor is electrical potential (ψ),

$$\frac{\partial (\Delta F^{\ddagger})}{\partial \psi} = \Delta q^{\ddagger} \tag{6.22}$$

6. Interface Effects on Chemical Reaction Rate

where $\Delta q^{\ddagger}$ is the difference in electrical charge between activated complex and reactant molecules. Then

$$\frac{\partial (\ln k)}{\partial \psi} = -\frac{\Delta q^{\ddagger}}{RT} \qquad (6.23)$$

Provided that q and ψ are of the same sign, the rate of reaction will therefore increase or decrease with increase of ψ, depending on whether $\Delta q^{\ddagger}$ is negative or positive, respectively.

In addition to these effects on the rate constant, interfacial pressure and electrical potential can influence the reaction rate by altering the distribution of species between bulk and interface according to Eq. (6.10). Large changes of reaction rate as a result of increased or reduced concentrations of reactive ions such as OH^- or H^+, caused by changes of the electrical potential at the interface, have been observed in a number of studies [35, 36, 37].

D. Effects of Accessibility of Reactive Groups

Reaction rates may be altered as a result of changes in accessibility of reactive groups caused by compression of the monolayer. For the oxidation of oleic acid monolayers by acid permanganate, present in the subphase, the vertical orientation of the hydrocarbon chains accompanying compression would tend to separate the mid-chain double bonds from the aqueous phase, thus making it more difficult for the permanganate ions to reach them. In fact, the rate constant for this reaction decreases with increasing interfacial pressure [38] as it does for a number of other interfacial reactions [39, 40, 41]. A mathematical theory has been developed by Mittelmann and Palmer [42] to explain this type of steric barrier to reactions at interfaces. From this theory, it is possible to calculate an accessibility factor (ϕ), which has a value of unity when the molecules all lie flat on the surface and decreases as the molecules tend to become vertically orientated. The rate constant for a surface reaction must therefore include the term ϕ, which varies with the interfacial pressure. Values of ϕ varied from 0.481 to 0.120 for the oxidation of triolein monolayers when the interfacial pressure was increased from 1 to 14 dynes/cm. Very good agreement was obtained between experiment and theory [42].

E. *Action of Inhibitors*

Interfacial reactions are, in some ways, more susceptible to inhibition than bulk reactions. Apart from the usual types of inhibitors which interfere with the reaction mechanism and which exert their effects in bulk solution, other types of inhibition, specific for interfaces, may operate. The adsorption of nonreactive tensioactive compounds may inhibit reactions at interfaces by simply preventing or reducing adsorption of the reactive species or by altering unfavorably the electrical potential at the interface. The build-up of products that do not desorb easily or precipitate in the interface (see Sect. 6.IV.C) is also a common source of inhibition. Mechanisms such as these are responsible for the poisoning and inhibition effects commonly observed in heterogeneous catalysis.

IV. REMOVAL OF PRODUCTS FROM INTERFACE

A. *Diffusion in Absence of Barrier to Desorption*

When the chemical potential of molecules in the adsorbed state becomes greater than in either of the adjacent bulk phases, desorption proceeds until chemical potentials are equalized. If there is no activation energy to the desorption step, then an equilibrium between interface and sublayer may be assumed (see Sect. 6.II.A). Under these conditions, the rate of desorption will be equal to the rate of diffusion from the sublayer into the bulk. If the concentration in the bulk phase is initially zero and the interfacial film is maintained at constant interfacial pressure, the rate of desorption is given by an equation identical to Eq. (6.3), where C_o is now the concentration of the desorbing substance in the sublayer, which is equivalent to

6. Interface Effects on Chemical Reaction Rate

the equilibrium bulk concentration for that value of the interfacial pressure. Equation (6.3) has been verified for the desorption of lauric acid from an air-water interface [12]. After desorption has proceeded for a given time, the extent of the concentration gradient near the interface becomes equal to the thickness of the stationary layer (see Sect. 6.II.A). At this point, the desorption kinetics change, since now solute that reaches the edge of the stationary layer is rapidly dispersed by convection; the rate of desorption then becomes constant and equal to the rate of diffusion through a layer of thickness x, one edge of which is, initially at least, effectively at zero concentration:

$$\frac{dn}{dt} = \frac{DC_0}{x} \tag{6.24}$$

Desorption has been shown to reach a steady state obeying Eq. (6.24) after a given time [12].

As for adsorption, the rate of desorption depends on the interfacial pressure and, if the molecule is charged, on the electrical potential near the interface. A simple way of reducing the rate of desorption of a charged species is to incorporate, in the monolayer, molecules carrying either no charge or a charge of opposite sign. It has been suggested that the dramatic effects on emulsion stability of adding nonionic surfactants to systems containing ionic surfactants [43] is due to a reduction in the rate of desorption of the ionic species [44, 45].

B. Diffusion in Presence of Barrier to Desorption

The presence of an activation energy barrier to desorption means that the equilibrium concentration at a given interfacial pressure will not be attained in the sublayer. Under these conditions, the rate of transfer of solute from the interfacial region is still governed by diffusion so that a $t^{\frac{1}{2}}$ relation is closely followed over a certain range [46, 47]. If the rate of desorption (dn/dt) is measured under steady state conditions at a given value of the interfacial pressure, and the equilibrium sublayer concentration (C_o) is known at this pressure from equilibrium measurements, the magnitude of the barrier to desorption, expressed as an interfacial resistance (R_2), may be calculated [48] from

$$\frac{dn}{dt} = \frac{C_o}{R_1 + R_2} \qquad (6.25)$$

where R_1 is the diffusional resistance and is equal to x/D, x being the thickness of the stationary layer and D the diffusion coefficient of the substance. No desorption barriers have been reported for simple molecules, but there is evidence of barriers to desorption for polymers. An interfacial resistance of 2.4×10^8 sec/cm was found for the desorption of the protein BSA from an air-water interface at 25.6 dynes/cm pressure [48].

C. Interfacial Precipitation

An interface, as mentioned earlier, is a two-dimensional solvent and every compound has a well-defined solubility at a given interface. It is customary to express this solubility limit in terms of an equilibrium spreading pressure [49], i.e., the interfacial pressure at which there is equilibrium between a compound in its bulk and monolayer states. If, for any reason, this solubility limit is exceeded, monolayer substance will tend to crystallize or condense to the bulk material. In some cases, supersaturated monolayers may remain stable for long periods, although precipitation occurs rapidly once their interfacial pressure reaches a critical value [50].

Polymerization in interfacial films presents some interesting points in connection with precipitation at interfaces. As an example, polyamides are formed at an interface if an aqueous solution of a diamine is brought into contact with a solution of a diacid chloride in carbon tetrachloride. It is known that this type of polymer becomes practically insoluble in either of the bulk phases above a comparatively low molecular weight, yet high molecular weights are obtained. This poses the question of how these high molecular weight polymers are obtained in a solution phase since precipitation of the polymer would be expected to occur before the polymerization had proceeded to any extent. The problem may be explained if we consider that the polymerization occurs in a monolayer at the interface, for here the solubility of the polymer is very high, corresponding to an interfacial pressure of about 14 dynes/cm [51]. Once this pressure is exceeded, however, precipitation of polymer occurs and may lead to the formation of a thick polymer skin at the interface [52]. Eventually, the film effectively eliminates all the interface so that the polymerization reaction ceases. Only by removing the precipitate as it forms can the reaction be made to continue.

V. EXPERIMENTAL APPROACHES TO MEASUREMENTS OF RATES OF REACTION

In general, a chemical reaction in a monolayer at a liquid-liquid or liquid-gas interface is accompanied by changes in molecular area, interfacial pressure, interfacial potential, electrical dipole moment, and interfacial viscosity or elasticity. The kinetics of reactions may therefore be followed by measuring the changes in one or more of these parameters. In addition, important supplementary information may be obtained by removing the products from the interface and subjecting them to chemical or physical analyses [53]. At interfaces with solids, techniques for following reactions are more limited and it is often necessary to examine the course of an interfacial reaction by monitoring the bulk phase. Change of ζ potential with time is sometimes a useful approach for following kinetics of a reaction.

The techniques for following interfacial reactions are adequately described in standard references [54, 55]. Here, only the planning and interpretation of experiments will be considered. For reactions in bulk, it is usually relatively simple to study reactions while the composition of the system remains constant. On the other hand, interfaces cannot easily be isolated. As a result, exchange of reactants and products with the bulk phases is often superimposed on the interfacial reaction. In order to interpret chemical reactions, it is therefore most important to have a sound knowledge of the kinetics of physical processes that are likely to occur at interfaces.

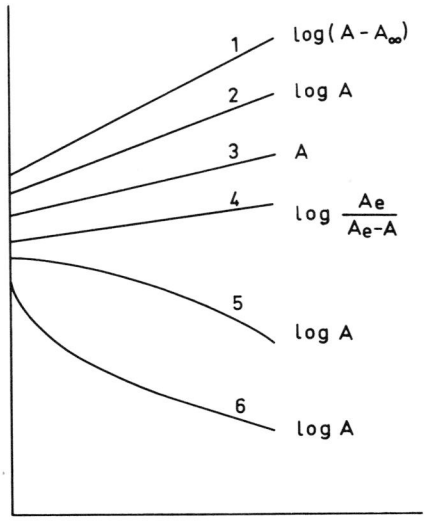

FIG. 6.3. Kinetics of different interfacial processes at constant interfacial pressure. (1) First-order chemical reaction. (2) Adsorption. (3) Spreading from bulk source of fixed perimeter. (4) Reversible expulsion of segments from a polymer monolayer. (5) Nucleation of a supersaturated monolayer. (6) Desorption.

For studying interfacial processes under as controlled conditions as possible, it is preferable to carry out measurements at constant interfacial pressure. The kinetics of several isolated interfacial processes are summarized below (see also Fig. 6.3). For a given interfacial reaction, one or more of these physical processes may be operating.

1. *Adsorption*: The rate of adsorption into a monolayer (Π, ψ constant) is equal to $(1/A)(dA/dt)n$ where A and dA/dt are the total area and rate of increase of area at a given time t and n is the concentration of molecules per unit area in the monolayer. The rate of adsorption may be evaluated from the slope of the graph of log A vs. t. If adsorption occurs at constant area, the rate of adsorption, dn/dt, is equal to $(d\Pi/dt)(dn/d\Pi)$ and may be determined

6. *Interface Effects on Chemical Reaction Rate* 129

provided that the n-Π relationship is known. In many cases, the n-Π relationship is linear over a wide range. Making the substitutions $\alpha = dn/d\Pi$ and $\beta = A/kt$, Eq. (6.7) can now be integrated to give an adsorption equation in which Π is a linear function of $\ln t$:

$$\Pi = \frac{1}{\beta} \ln t + \frac{1}{\beta} \ln \frac{k_1 C_o}{\alpha} + \text{const.} \qquad (6.26)$$

2. *Desorption*: The rate of desorption (Π, ψ constant) varies with $t^{-\frac{1}{2}}$ in the initial stages and with t in the later stages. For desorption from a monolayer at constant area per molecule, the kinetics follow an equation [56, 57] of the form

$$2.3 \left(\frac{\partial (\log A_f)}{\partial t^{\frac{1}{2}}}\right)_A = -2K_d \left(\frac{D}{\pi}\right)^{\frac{1}{2}} \qquad (6.27)$$

where A_f is the area of monolayer, K_d the equilibrium ratio of bulk to surface concentration, and D the diffusion coefficient.

3. *Spreading of monolayer from bulk crystal or liquid drop at the interface*: For a spreading source whose perimeter does not change with time, A increases linearly with t (Π constant).

4. *Precipitation at interface*: In the case of spontaneous nucleation of a supersaturated monolayer, the rate $[(1/A)(dA/dt)n]$ increases with time, at least in the initial stages [58]. For precipitation on a fixed perimeter, A decreases linearly with t.

5. *First-order reversible process at interface such as the expulsion and reentry of polymer segments* [59]: If A_e is the area at equilibrium and A the area at any time t, $\log [A_e/(A_e - A)]$ is a linear function of t.

Turning now to chemical reactions, the general expression for a reaction obeying first-order kinetics is:

$$n = n_o e^{-kt} \qquad (6.28)$$

where n_o is the initial concentration of reactant and n is the concentration at time t. If, in an interface reaction, A_o is the initial area of the monolayer, A_∞ the area after completion of the reaction (i.e., the area occupied by product), and A the area at any time t then, assuming additivity of molecular areas of product and reactant, replacement of molar concentrations by areas in Eq. (6.28) leads to the expression

$$\frac{A - A_\infty}{A_o - A_\infty} = e^{-kt} \qquad (6.29)$$

A plot of log $(A - A_\infty)$ vs. t should be a straight line, from which k, the rate constant, may be evaluated (Fig. 6.3). Any other additive interfacial parameter such as interfacial potential may be used in place of area in Eq. (6.29). Several equations, applicable to certain types of first-order interfacial reactions, are given by Adamson [60].

Corresponding expressions for higher-order chemical reactions may be similarly developed. In practice, however, it is probably better to consider each particular reaction individually; it is always necessary to make allowances for changes of areas of reactant and product molecules as the composition of the monolayer changes during reaction, as well as to take into account any superimposed physical processes. Even for first-order reactions some of the assumptions, including additivity of molecular areas, are usually not valid [61]. In mixed monolayers of several components, it is useful to follow reactions using simultaneous measurements such as area, interfacial potential, and interfacial viscosity. Then, using previously obtained calibration curves for monolayers of various compositions, it is possible to determine the composition of the monolayer at different stages of the reaction, where this would have been impossible from area measurements alone. Often, quantitative measurements of reaction rates cannot be made, yet much useful information may be gained by interfacial measurements. For example, polymers generally give highly viscous films, so that polymerization may often be detected by interfacial viscosity measurements [51].

VI. EXPERIMENTAL RESULTS

Based on the considerations of Sect. 6.III, it would be expected that reactions at interfaces would, in many instances, be much more

6. Interface Effects on Chemical Reaction Rate

rapid than the same reactions in bulk. Measurements, by the classical surface chemistry approach, have revealed, however, that the rate constants and activation energies are similar for bulk and interface reactions [61]. The reactions studied have usually been between a compound spread as an insoluble monolayer and nontensioactive ions dissolved in the subphase, e.g., the hydrolysis of esters [40] and the oxidation of unsaturated fats and fatty acids [38, 42]. By comparison, little data is available on reactions between tensioactive compounds or between compounds spread as insoluble monolayers. These types of reaction are to be preferred for elucidating interfacial rate theory. In contrast, studies of interfacial polycondensation reactions [62] have revealed extremely fast rates in many cases. Assuming the reactions to be two-dimensional [51], some confirmation of the theory presented in Sect. 6.III is obtained. However, it is doubtful whether some of the rate constants reported are the true rate constants for the two-dimensional reaction, since precipitation of the polymer usually occurs and has a great retarding effect on the reaction [63]. The measurement of true rate constants presents great problems because of the speed of these reactions. A few selected examples of interface effects on reaction rates will now be considered.

A. Effect of Interfacial Pressure

The results of Gee [64] for the kinetics of polymerization of monolayers of the oxidized form of the maleic anhydride-β-elaeostearin adduct are of particular interest since the reactants are held in an insoluble monolayer. In Fig. 6.4 the logarithm of the rate constant is plotted against interfacial pressure for the data of Gee. Linear relationships are obtained, and the slopes are similar for three different experimental conditions. An average value of 138 $\overset{\circ}{A}{}^2$ is calculated for $\Delta A^{\ddagger}$. Since the rate increases with increasing interfacial pressure for this polymerization, $\Delta A^{\ddagger}$ should represent the decrease in area when the transition complex is formed from the reactants. In view of the overall increase in area accompanying

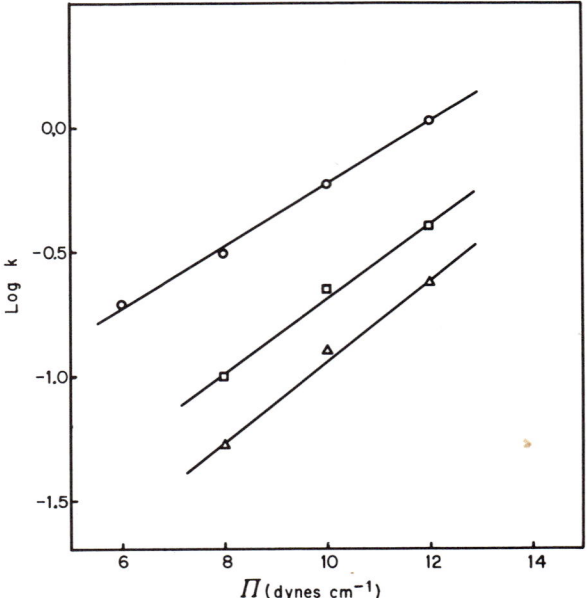

FIG. 6.4. Log (rate constant) vs. Π for polymerization of oxidized maleic anhydride-β-elaeostearin adduct.
$-\bigcirc- \frac{N}{100} H_2SO_4 + 2 \times 10^{-3}$ M $CoSO_4$, 303°K; $-\square- \frac{N}{100} H_2SO_4$, 308°K;
$-\triangle- \frac{N}{100} H_2SO_4$, 303°K.

the polymerization, the interpretation of $\Delta A^{\ddagger}$ is difficult and must await further results of a similar nature. In all reactions between insoluble monolayers and dissolved ions which have been studied, the rate of reaction is found to decrease markedly with increase of interfacial pressure. A true pressure effect on these reactions, as predicted by Eq. (6.21) has not been considered, although since most of these reactions involve an expansion of area at constant Π, when products are formed from reactants, such an effect might be important.

B. *Effect of Electrical Potential*

Llopis and Davies [65] studied the effect of electrical potential on the rate of hydrolysis of cholesterol formate by H^+ ions. The

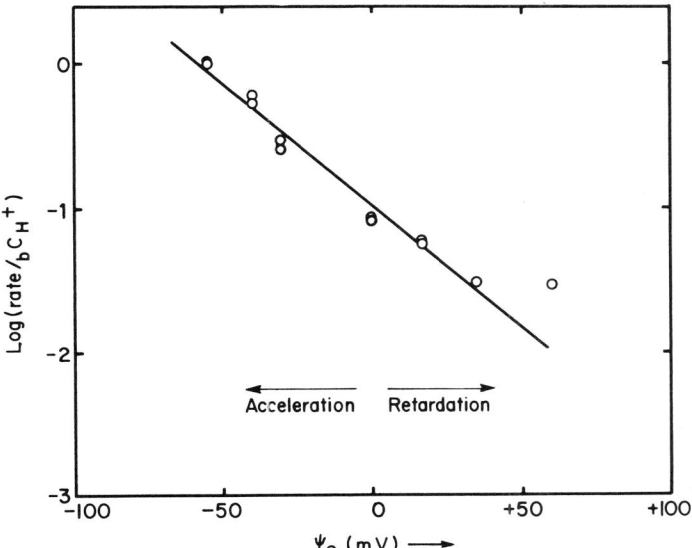

FIG. 6.5. Plot of log (rate/$_b C^{H^+}$) against ψ_G for the hydrolysis of cholesterol formate monolayers by HCl (from Ref. 54).

electrical potential in the monolayer was varied by incorporation of different amounts of a long-chain sulfate, $C_{22}H_{45}SO_4^-$, or a long-chain quaternary ammonium ion, $C_{18}H_{27}N(CH_3)_3^+$, into the ester film. The rate of this reaction is proportional to the H^+ concentration at the interface. Distribution of H^+ between bulk and interface should be given by Eq. (6.10) with all terms constant except the one containing ψ. When the logarithm of the rate divided by the bulk H^+ concentration was plotted against ψ, calculated using the theory of Gouy, a straight line was obtained (see Fig. 6.5) with a slope equal to the theoretical value of -60 mV predicted from Eq. (6.10).

C. *Effect of Accessibility of Reactants*

As mentioned in Sect. 6.III.D, the decrease of the rate constant with increasing interfacial pressure has been interpreted in terms of an accessibility factor, ϕ, which decreases as Π increases [42]. However,

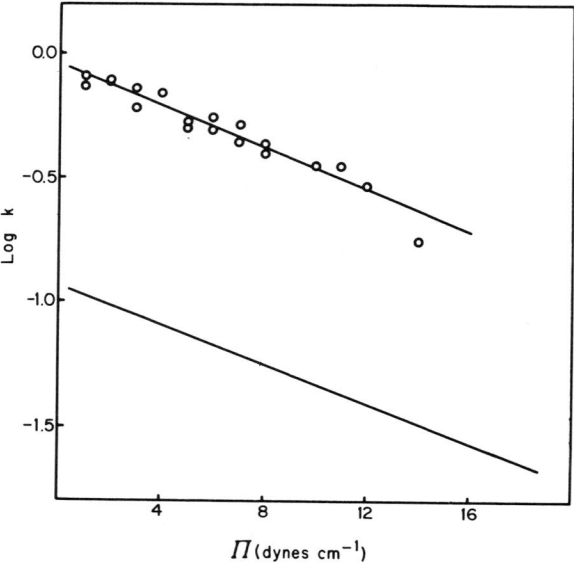

FIG. 6.6. Plots of log (rate constant) against interfacial pressure for oxidation of monolayers by permanganate. _O_O_O_ triolein (data of Mittelmann and Palmer); _____ oleic acid (data of Hughes and Rideal).

Eq. (6.10) provides an alternative explanation of this effect. The rate constants for reactions between insoluble monolayers and ions, dissolved in the bulk phase, are found to be directly proportional to the bulk concentration of ions at a fixed value of Π. The relevant concentration should, however, be that in the interfacial phase as confirmed by measurements in which ψ has been varied, keeping the bulk concentration constant. A rate proportional to the interfacial concentration is consistent with experiment since the distribution coefficient of an ion between bulk and interface should be approximately constant at a fixed value of Π. When Π is varied however, the distribution coefficient of the ion, as well as all other species, should change in accordance with Eq. (6.10). In Fig. 6.6 the logarithm of the rate constant is plotted against Π for the permanganate oxidation of triolein monolayers, using the results of Mittelmann

6. Interface Effects on Chemical Reaction Rate

and Palmer [42]. The correlation coefficient for fitting the points to a linear plot was very high and not significantly different to that for the rate constant vs. ϕ plot. Assuming k to be directly proportional to n_i, which in this case is the concentration of permanganate ions in the interfacial phase, application of Eq. (6.10) yields a value of 37 $\overset{\circ}{A}{}^2$ for A. This value is not inconsistent with the area expected for a permanganate ion in the interface. A plot of log k vs. Π from the data of Hughes and Rideal [38] for the oxidation of oleic acid monolayers is also shown in Fig. 6. The slope of the line is remarkably similar to that for triolein.

D. Reactions at Unidimensional and Zero-Dimensional Interfaces

Many examples have been reported of reactions that initiate at the linear interface between two solids or between a liquid and a solid and subsequently spread in two dimensions. It is found that reaction practically always begins at an edge, preferably at a corner, and rarely at a face. Reactions of this type, which may often be followed by using a microscope, include the reduction of copper oxide by hydrogen, the efflorescence of sodium carbonate, sulfate, or phosphate, the formation of red from yellow mercuric oxide, the escape of carbon dioxide from carbonates, and the decomposition of silver oxalate, selenite crystals, and permanganates [for further discussion see Refs. 66 and 67]. All these reactions illustrate the great specificity of unidimensional and zero-dimensional interfaces as reaction sites.

Since the pioneering work of Rideal and his school at Cambridge (extended by Davies) and of Havinga and his school at Leiden, the study of reactions at interfaces, by classical surface chemistry methods, has been comparatively neglected (see, however, Refs. 68, 69, 70 for some recent studies). This is unfortunate in view of the great importance of this field. Many facets of basic theory require reinvestigation, while new and fascinating avenues of research await workers in the field. One obvious area that is likely to expand in activity is that concerned with polymerization at interfaces. In nature, polymers of great specificity and homogeneity are synthesized without the need for resorting to extreme conditions of temperature and pressure. Attempts to understand the principles involved in biological synthesis and to simulate them in order to produce synthetic analogs or polymers having completely novel properties offer exciting prospects. Another aspect of interest would be to test to what extent enzyme reactions can be explained as interfacial reactions. This would require evaluation of partition functions for transition complexes and reactants from thermodynamic data. Studies along these lines might reveal interesting information about enzyme reactions.

REFERENCES

1. A. W. Adamson, *The Physical Chemistry of Surfaces,* Interscience, New York, 1960, pp. 143-152, 284-288.
2. G. L. Gaines, *Insoluble Monolayers at Liquid Gas Interfaces,* Wiley-Interscience, New York, 1966, Chap. 7.
3. J. T. Davies, *Advances in Catalysis, 6*:1 (1954).
4. J. F. Danielli and J. T. Davies, *Advances in Enzymology, 9*:35 (1951).
5. J. H. deBoer, *The Dynamical Character of Adsorption,* Clarendon, Oxford, 1953, p. 30.
6. J. Crank, *The Mathematics of Diffusion,* Clarendon, Oxford, 1956, p. 34.
7. F. MacRitchie and A. E. Alexander, *J. Colloid Sci., 18*:453 (1963).
8. I. Langmuir and V. J. Schaefer, *J. Amer. Chem. Soc., 59*:2400 (1937).
9. V. G. Levich, *Physicochemical Hydrodynamics,* Scripta Technica, Trans. Prentice-Hall, Englewood Cliffs, N.J., 1963.
10. J. M. Coulson and J. F. Richardson, *Chemical Engineering,* Vol. 1, Pergamon, London, 1955, p. 264.
11. F. MacRitchie, *Nature, 218*:669 (1968).
12. L. Ter Minassian-Saraga, *J. Chim. Phys., 52*:181 (1955).
13. F. MacRitchie, *Science, 163*:929 (1969).
14. A. F. H. Ward and L. Tordai, *J. Chem. Phys., 14*:453 (1946).

15. R. Defay and J. R. Hommelen, *J. Colloid Sci.*, *14*:411 (1959).
16. A. M. Posner and A. E. Alexander, *Trans. Faraday Soc.*, *45*:651 (1949).
17. A. F. H. Ward and L. Tordai, *Rec. Trav. Chim. Pays-Bas.*, *71*:572 (1952).
18. I. Langmuir, *J. Amer. Chem. Soc.*, *39*:1848 (1917).
19. F. MacRitchie and A. E. Alexander, *J. Colloid Sci.*, *18*:458 (1963).
20. A. Silberberg, *J. Phys. Chem.*, *66*:1872 (1962).
21. F. MacRitchie, *J. Macromol. Sci.-Chem.*, *A4*:1169 (1970).
22. G. Gouy, *J. Phys. Radium*, *9*:457 (1910).
23. O. Stern, *Z. Electrochem.*, *30*:508 (1924).
24. J. T. Davies, *Trans. Faraday Soc.*, *48*:1052 (1952).
25. G. S. Hartley and J. W. Roe, *Trans. Faraday Soc.*, *36*:101 (1940).
26. F. MacRitchie and A. E. Alexander, *J. Colloid Sci.*, *18*:464 (1963).
27. K. Imahori, *Bull. Chem. Soc. Jap.*, *25*:13 (1952)(in English).
28. A. G. Tweet, W. D. Bellamy, and G. L. Gaines, Jr., *J. Chem. Phys.*, *41*:2068 (1964).
29. I. Langmuir, *Colloid Symposium Monograph,* Chemical Catalog, New York, 1925, p. 48.
30. A. M. Posner, J. R. Anderson, and A. E. Alexander, *J. Colloid Sci.*, *7*:623 (1952).
31. G. Jura, *J. Phys. Colloid Chem.*, *52*:40 (1948).
32. I. Prigogine and L. Saraga, *J. Chim. Phys.*, *49*:399 (1952).
33. W. H. Carothers, *Trans. Faraday Soc.*, *32*:39 (1936).
34. S. Glasstone, K. J. Laidler, and H. Eyring, *The Theory of Rate Processes,* McGraw-Hill, New York, 1941.
35. A. E. Alexander and E. K. Rideal, *Proc. Roy. Soc., Ser. A*, *163*:70 (1937).
36. E. Havinga, in *Monomolecular Layers* (H. Sobotka, ed.), A. A. A. S, Washington, 1954, p. 192.
37. J. T. Davies and E. K. Rideal, *Proc. Roy. Soc., Ser. A*, *194*:417 (1948).
38. A. H. Hughes and E. K. Rideal, *Proc. Roy. Soc., Ser. A*, *140*:253 (1933).
39. J. Marsden and E. K. Rideal, *J. Chem. Soc.*, 1163 (1938).
40. A. E. Alexander and J. H. Schulman, *Proc. Roy. Soc., Ser. A*, *161*:115 (1937).
41. J. T. Davies, *Trans. Faraday Soc.*, *45*:448 (1949).

42. R. Mittelmann and R. C. Palmer, *Trans. Faraday Soc.*, *38*:506 (1942).
43. J. H. Schulman and E. G. Cockbain, *Trans. Faraday Soc.*, *36*:651 (1940).
44. Ref. 1, p. 394.
45. F. MacRitchie, *Nature*, *215*:1159 (1967).
46. Ref. 6, p. 38.
47. J. T. Davies and J. B. Wiggill, *Proc. Roy. Soc., Ser. A*, *225*:277 (1960).
48. G. Gonzalez and F. MacRitchie, *J. Colloid Interfac. Sci.*, *32*:55 (1970).
49. A. Cary and E. K. Rideal, *Proc. Roy. Soc., Ser. A*, *109*:301, 318, 331 (1925).
50. N. K. Adam, *The Physics and Chemistry of Surfaces*, Oxford, 1938, p. 104.
51. F. MacRitchie, *Trans. Faraday Soc.*, *65*:2503 (1969).
52. F. MacRitchie and N. F. Owens, *J. Colloid Interfac. Sci.*, *29*:66 (1969).
53. F. Kögl and E. Havinga, *Rec. Trav. Chim. Pays-Bas*, *59*:249, 323, 601 (1940).
54. J. T. Davies and E. K. Rideal, *Interfacial Phenomena*, Academic, New York, 1961.
55. A. E. Alexander and G. E. Hibberd, in *Physical Methods of Chemistry* (A. Weissberger and B. W. Rossiter, eds.), Wiley, 1971.
56. L. Ter Minassian-Saraga, *J. Colloid Sci.*, *11*:398 (1956).
57. J. H. Brooks and B. A. Pethica, *Trans. Faraday Soc.*, *60*:208 (1964).
58. J. H. Brooks and A. E. Alexander, in *Retardation of Evaporation by Monolayers* (V. K. La Mer, ed.), Academic, New York, 1962, p. 245.
59. F. MacRitchie, *J. Colloid Sci.*, *18*:555 (1963).
60, Ref. 1, p. 144.
61. A. R. Gilby and A. E. Alexander, *Aust. J. Chem.*, *9*:347 (1956).
62. P. W. Morgan, *Condensation Polymers by Interfacial and Solution Methods*, Wiley, New York, 1965.
63. P. W. Morgan and S. L. Kwolek, *J. Polymer Sci.*, *62*:33 (1962).
64. G. Gee, *Proc. Roy. Soc., Ser. A*, *153*:129 (1935).
65. J. Llopis and J. T. Davies, *An. Real Soc. Espan. Fís. Quím.*, *49*:671 (1953).

66. Ref. 1, p. 285.
67. W. D. Harkins, *The Physical Chemistry of Surface Films*, Reinhold, New York, 1952, p. 291.
68. I. R. Miller and J. M. Ruysschaert, *J. Colloid Interfac. Sci.*, *35*:340 (1971).
69. M. Iwahashi, *J. Colloid Interfac. Sci.*, *50*:532 (1975).
70. A. Dubault, C. Casagrande, and M. Veyssie, *J. Phys. Chem.*, *79*:2254 (1975).

7.

LIQUID-VAPOR INTERFACIAL POLYCONDENSATION

L. B. Sokolov

Department of Polycondensation Processes
All-Union Research Institute of Synthetic Resins
Vladimir, USSR

I.	PROCEDURE OF POLYCONDENSATION PROCESS	142
II.	SYNTHESIS OF VARIOUS POLYMERS	143
III.	SYNTHESIS OF COPOLYMERS	145
IV.	EFFECT OF SOME PARAMETERS ON THE MOLECULAR WEIGHT AND YIELD OF POLYMERS	149
V.	FEATURES OF THE COURSE OF LIQUID-VAPOR INTERFACIAL POLYCONDENSATION	154
VI.	CONCLUSION	164
	REFERENCES	165

The technique of liquid-vapor (liquid-gas in Soviet literature) interfacial polycondensation was established by the author and his co-workers in 1960 [1].

This chapter, in addition to previously published papers [2] and monographs [3, 4], pertains to the main relationships of the liquid-vapor polycondensation.

I. PROCEDURE OF POLYCONDENSATION PROCESS

Liquid-vapor interfacial polycondensation utilizes pairs of highly-reactive monomers with one of the monomers in the vapor (gaseous) state and the other in solution. In the synthesis of polyamides, the polycondensation reaction takes place at the interface between the aqueous solution of the diamine and the vapor of the diacid chloride. Since diacid chlorides are more volatile than diamines they are preferentially employed in the vapor phase, usually mixed with an inert gas such as nitrogen. The solvent for diamines is usually water since it readily dissolves the amine hydrochloride formed although use of organic solvents is not excluded.

The contact of the monomers in liquid-vapor polycondensation is readily effected by bubbling the vapor (gaseous) monomers through the aqueous solution of the other monomer. In order to prevent the possibility of condensation of the reagent vapor on apparatus walls, the vapor mixture is overheated or is diluted so that the partial pressure of the vapor reagent in the gas mixture is lower than its saturation vapor pressure.

In liquid-vapor interfacial polycondensations, a polymer is formed at once producing a film covering on a gas bubble passing through the liquid layer. The polymer film is separated due to the efflux of the gaseous reagent bubble from the liquid. If the polymer obtained is of low molecular weight or is incapable of flexible film formation a fine-dispersed powder will be formed.

The most appropriate conditions for the synthesis of a great many high-molecular-weight polymers by the liquid-vapor interfacial polycondensation are as follows: reaction temperature about $95°C$; monomer concentration in the aqueous phase about 0.1 ± 0.05 mol/liter monomer concentration in the vapor phase not more than 50% by volume; height of the solution layer not less than 50 mm. The duration of an experiment may vary but should not be so great as to effect a considerable decrease in the monomer concentration of the liquid phase. After a definite time of reaction (usually 0.5-1.5 hr) the polymer is filtered from the liquid phase, washed, and dried.

II. SYNTHESIS OF VARIOUS POLYMERS

Table 7.1 comprises a list of polymers obtained by liquid-vapor interfacial polycondensations [3]. Data on the syntheses of these polymers in systems of two immiscible liquids are also given for comparison.

It is readily seen that a wide range of polymers may be prepared by liquid-vapor interfacial polycondensation. It follows from Table 7.1 that one or both of the monomers used in a liquid-vapor interfacial polycondensation are rather reactive. For instance, the preparation of polyamides from active diamines and easily vaporized dichlorides of strong dicarboxylic acids are the most successful. Other polymers may also be produced by this method if the polycondensation reaction is very rapid. The polycondensation technique proves to be most effective if the more reactive monomers are situated in the vapor phase, even if they are the more easily hydrolyzable reactants (see section 7.V). If a reagent, metered in the vapor phase, is of relatively low reactivity and not easily hydrolyzable, then both the liquid-vapor and liquid-liquid techniques result in similar products being formed.

Special consideration should be given to the synthesis of polyamides from diacid chlorides of fluorinated dicarboxylic acids. Introduction of fluorine atoms into a dicarboxylic acid results in a sharp increase in its acidic properties and lowers the boiling point of the acid and its derivates. The use of such reactive fluorinated acid chlorides in liquid-liquid interfacial polycondensation has been discouraging, but the same monomers may be used successfully in liquid-vapor polycondensation. Thus, presently, liquid-vapor interfacial polycondensations have turned out to be the most effective polycondensations involving diacid chlorides of fluorinated dicarboxylic acids. (Note, in Table 7.1, 1,6-polyhexamethylene perfluoroadipamide by the liquid-vapor method.)

The foregoing also pertains to diacid fluorides of perfluoralkanedicarboxylic acids. Although, in general, acid fluorides of dicarboxylic acids are less active than acid chlorides in reactions

TABLE 7.1

Polymers Synthesized by Liquid-Vapor Interfacial Polycondensation

Monomer in vapor phase	Monomer in liquid (aqueous) phase	Reduced viscosity 0.5 g/gl polymer solution (dl/g)		Solvents*
		Liquid-vapor system	Liquid-liquid system	
Polyoxamides				
Oxalyl fluoride	1,6-Hexamethylenediamine	0.80		H_2SO_4
Oxalyl chloride	1,6-Hexamethylenediamine	1.50	0.45	H_2SO_4
	1,10-Decamethylenediamine	0.64	—	H_2SO_4
	1,4-Phenylenediamine	2.12	0.42	HSO_3Cl
	1,3-Phenylenediamine	0.30	—	H_2SO_4
Fluorinated polyamides, etc.	Benzidine	2.50		HSO_3Cl
Perfluoroadipoyl chloride	1,6-Hexamethylenediamine	0.53	0.24	H_2SO_4
	Ethylenediamine	0.12	—	H_2SO_4
	1,4-Phenylenediamine	0.64	0.09	HSO_3Cl
	1,3-Phenylenediamine	0.46	—	H_2SO_4
Carbon suboxide	1,6-Hexamethylenediamine	1.10	0.80	H_2SO_4
Polyureas and Polythioureas				
Phosgene	1,6-Hexamethylenediamine	1.02	1.15	H_2SO_4
Thiophosgene	1,6-Hexamethylenediamine	0.50	—	1,3 cresol
	1,4-Phenylenediamine	0.31	0.76	1,3 cresol
Polythioesters				
Oxalyl chloride	1,4-Tetramethylenedithiol	nonsoluble	—	
	1,5-Pentamethylenedithiol	0.53	—	Tetraclorethane

Source: Ref. 3.

*For viscosity measurements.

7. Liquid-Vapor Interfacial Polycondensation

with amines, the acid fluorides of perfluoroalkanedicarboxylic acid nevertheless may be used.

In some cases attempted liquid-vapor interfacial polycondensations did not result in the formation of high-molecular-weight polymers. Reasons for limitations against achieving high molecular weight of a polymer during its synthesis in liquid-vapor systems are as follows: hydrolysis of the polymer formed (e.g., in the synthesis of aromatic polyesters); low rate of the propagation reaction (e.g., in the system of 1,4-phenylene-diamine-phosgene); and formation of cyclic products (e.g., during polycondensation of ethylendiamine with oxalyl chloride).

III. SYNTHESIS OF COPOLYMERS

The synthesis of copolymers in liquid-vapor systems presents no difficulties beyond those in the synthesis of homopolymers. The difference lies only in using a mixture of two or more monomers in the aqueous or the vapor phase.

In liquid-vapor interfacial polycondensations the polymer composition is dependent on the composition of the initial mixture of monomers (both in liquid and vapor phases). Figure 7.1 shows an example of such a dependence. As a rule, one can see a great difference between the copolymer composition and that of the initial mixture of monomers. Therefore, it is necessary to take into consideration the quantitative ratio between the copolymer composition and that of the initial mixture of reagents to prepare a copolymer of the definite composition. For polycondensation, for instance, of two diamines A and B with one diacid chloride C, the relation between the copolymer composition and that of the initial diamine mixture during liquid-vapor interfacial polycondensation may be expressed as follows:

$$\frac{a}{b} = r \frac{[A_o]}{[B_o]} \qquad (7.1)$$

where a/b is the relation of contents of monomer units A and B in the copolymer; A_o, B_o are the initial concentrations of monomers

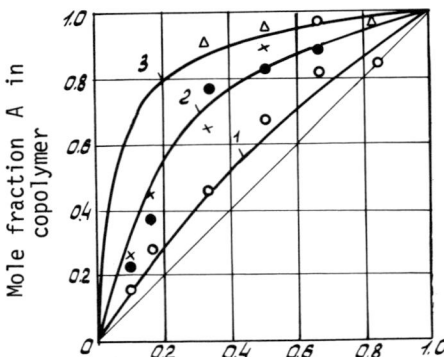

Mole fraction A
in initial mixture of diamines

FIG. 7.1. Dependence of copolymer composition on the composition of the initial mixture of monomers in the polycondensation of two aliphatic diamines with oxalyl chloride and phosgene at the liquid-vapor system [3]. Relative diamine mole fraction, A = (A)/(A + B).
1. 1,6-Hexamethylenediamine (A), ethylenediamine (B), oxalyl chloride. 2. 10-Decamethylene diamine (A), 1.6-hexamethylenediamine (B), oxalyl chloride (o-based on nitrogen content; x-based on carbon content). 3. 1,10-Decamethylenediamine (A), ethylenediamine (B), phosgene.

(diamines) A and B in the liquid (aqueous) phase; and r is the experimental parameter of copolycondensation showing how much faster monomer A is copolycondensed than monomer B at present conditions. In the simplest case (homogenous system, slow polycondensation reactions) $r = r_o = k_1/k_2$, where k_1 and k_2 are the constants of the rate of chemical reaction of monomers A and B with intermonomer C, respectively.

In more difficult cases, which also include liquid-vapor polycondensation systems, as a rule, $r \neq r_o$; e.g., r includes values characteristic of surface activity of monomers, etc. (see below and also Chap. 8).

Equation (7.1) is valid for low degrees of conversion, i.e., below 5%. Its other form must be used for higher degrees of conversion:

7. Liquid-Vapor Interfacial Polycondensation

$$\log \frac{[A]}{[A_o]} = r \log \frac{[B]}{[B_o]} \qquad (7.2)$$

where A, B are the existing concentrations of monomers A and B in a reaction system at a specific time.

Table 7.2 gives experimentally determined values of the copolycondensation parameters for several liquid-vapor systems [3, 5]. It is obvious that since the copolycondensations, as a rule, are accompanied by other processes the copolymer composition will be dependent not only on the relative chemical activities but also on other factors, such as the surface activities of reagents, their susceptibility to hydrolysis, etc.

For liquid-vapor interfacial polycondensation r can be related to r_o as

$$r = r_o \cdot f \; (L, V_h, \ldots) \qquad (7.3)$$

where L is the relative surface activity of monomers and V_h is relative rate of hydrolysis. It was found [5] in the copolycondensation of two surface active diamines (e.g., two linear aliphatic diamines) that r can be related to r_o as follows:

$$r = r_o L^{n_A - n_B} \qquad (7.4)$$

where n_A and n_B are the number of $-CH_2-$groups in the diamines, and L is Traube's coefficient. L is equal to a quotient of specific surface activities of adjacent terms within a homologous series. For aliphatic monomers $L = e^{\Delta E/RT}$, where ΔE is energy released when one of the $-CH_2-$groups migrates from the liquid phase to the surface [5]. The value of ΔE experimentally obtained from the copolymer composition in liquid-vapor interfacial polycondensations is 346 cal/mol per methylene group. The same energy value is found from direct surface tension measurements (i.e., 354 cal/mol) [6].

The above indicates that the surface activities of aliphatic diamines are of great importance in copolycondensations in liquid-vapor systems. This confirms the view that the process of formation

TABLE 7.2

Copolycondensation Parameter r of Various Pairs
of Diamines in Aqueous Liquid-Vapor Systems (at 95°C)

Diamines		Diacid chloride [C]	r
A	B		
1,10-Decamethylenediamine	Ethylenediamine	Phosgene	14.3
	1,4-Tetramethylenediamine	Oxalyl chloride	9.0
	1,6-Hexamethylenediamine		4.8
1,6-Hexamethylenediamine	Ethylenediamine	Phosgene	5.0
		Oxalyl chloride	1.6
1,8-Octamethylenediamine	1,6-Hexamethylenediamine	Phosgene	1.7
Benzidine	1,4-Phenylenediamine	Oxalyl chloride	2.2
	1,6-Hexamethylenediamine		1.2
			4.3

Sources: Refs. 3 and 5.

7. *Liquid-Vapor Interfacial Polycondensation*

of polymers and copolymers proceeds just at the interface of liquid-vapor in the liquid-vapor polycondensation.

The copolymer composition will also be determined by the relative susceptibilities to hydrolysis when using two diacid halides in their gaseous form for synthesis of copolyamides. A study of the polycondensation of 1,6-hexamethylenediamine with a mixture of phosgene and oxalyl chloride has established that the copolymer content of oxalyl chloride (the most easily hydrolyzable reagent) is lower than could be expected on the basis of its higher reactivity in comparison to phosgene. The relative hydrolysis rates and reactivities are the main factors determinating copolymer composition since the surface activity of phosgene and oxalyl chloride is not high.

IV. EFFECT OF SOME PARAMETERS ON THE MOLECULAR WEIGHT AND YIELD OF POLYMERS

The main relationships in liquid-vapor interfacial polycondensations of the effects of reaction temperature and reagents concentration, pH of the aqueous phase, etc. on the molecular weight and yield of a polymer have been previously studied and described in detail [2, 3]. Table 7.3 gives a comparison of some of these dependences with systems containing two immiscible liquids.

It is seen from Table 7.3 that some dependences in liquid-vapor interfacial polycondensations are analogous to the dependences in two-phase liquid systems, but also that some do differ. The similarities of polycondensation processes in liquid-liquid and liquid-vapor systems is based on the heterogeneity of the systems; specific differences of these processes result from the more expressed surface character of reactions in liquid-vapor systems and from features of the vapor state of the gaseous reagents.

The most distinguishing feature of liquid-vapor interfacial polycondensations is a dependence of polymer molecular weight on the temperature in the liquid-vapor system: molecular weight increases with increase in the reaction temperature. During a polycondensation

TABLE 7.3

Comparison of Dependences of
Molecular Weights of Some Aliphatic
Polyamides on Synthesis Parameters

Reaction variable	Molecular weight change in systems	
	Liquid-liquid	Liquid-vapor
Increase in reaction temperature	Decreases	Increases
Addition of emulgators	Decreases	No variation
Increase of alkali content in an aqueous phase	Decreases	Decreases
Increase of diamine concentration in an aqueous phase	Maximum	Maximum

in a system of two immiscible liquids the molecular weight decreases with increasing reaction temperature (at temperatures above 30°-50°C); such dependences are often experimentally observed [6a].

The character of the temperature dependence can be combined with specific features of the gaseous state of one of the monomers. In fact, if the reaction of polymer formation is known to occur at the interface and hydrolysis of a gaseous monomer takes place in the aqueous phase, the dependence of the ratio between the main reaction rate V_p and the hydrolysis rate V_h with the temperature may be expressed as follows [3]:

$$\frac{V_p}{V_h} = Be^{-[E_p-(E_h-Q)/RT]} \tag{7.5}$$

where E_p, E_h are the activation energies of the propagation reaction and hydrolysis, respectively; Q is the heat of solution of a gaseous reagent in the aqueous phase; and B is a constant. The relationship V_p/V_h describes the dependence of the molecular weight of the polymer on the temperature.

Equation (7.5) as applied to polycondensation in systems of two immiscible liquids can be transformed into:

7. Liquid-Vapor Interfacial Polycondensation

$$\frac{V_p}{V_h} = Be^{-[E_p-(E_h+Q)/RT]} \tag{7.6}$$

It is evident from a comparison of Eqs. (7.5) and (7.6) that they differ in the sign of the heat of solution, Q, which expresses the different temperature dependences of liquid and gas solubilities in liquids.

E_p was estimated using Eq. (7.5) to be about 5 kcal/mol for a system of 1,6-hexamethylenediamine-phosgene. This value is reasonable and agrees with data in Ref. 7 on reactions of aniline with benzoyl chloride in nitrobenzene (E = 6.3 kcal/mol).*

These equations predict that molecular weight of the polymer during polycondensation in a liquid-vapor system will be increased with increasing reaction temperature. From these facts it is seen that the temperature dependence of polymer molecular weight can be a diagnostic procedure for determining the nature of process details in liquid-vapor systems.

A decrease in the molecular weight with increasing temperature is evidence that polycondensation is occurring further in the aqueous phase. An example of such nature of the heterophasic process is liquid-vapor polycondensation of the system 1,4-phenylenediamine (water)-phosgene (vapor). This system is characterized by a low rate of reaction, a fairly high (10 g/liter atm) solubility of the gaseous reagent (phosgene), and a relatively low susceptibility of phosgene to hydrolysis. The intrinsic viscosity of the resulting polymer is 0.14 dl/g 0°C. and 0.08 dl/g 90°C. Equations (7.5) and (7.6) should only be considered as semiquantitative and are used here to illustrate that the above point of view on the fundamental difference in the

---

*The activation energy, Ep, for propagation in polyamidations may be estimated from the data of L. V. Kuritsyn (7) for the analogous reaction of aniline with benzoyl peroxide: 4.3 (in anisole), 5.7 (acetone), and 6.3 kcal/mol (nitrobenzene). Initial values for this calculation were: for E_h, the activation energy for hydrolysis of acid chlorides, 12 kcal/mol (c.f. *Tables of Chemical Kinetics*, U.S. Department of Commerce, National Bureau of Standards, Washington, 1951); for heat of solution, Q, of phosgene in water, 6.8 kcal/mol (c.f. W. H. Manogue and P. L. Pigford, *A.I.Ch.E.J.*, 6:494 (1960).

temperature dependence of polymer molecular weight between liquid-liquid and liquid-vapor interfacial polymerization is true.

The dependences of the molecular weight and yield of polymer on the reagent phase concentrations during liquid-vapor interfacial polycondensation are expressed as curves with maxima. Figure 7.2 shows these dependences on oxalyl chloride concentration in the vapor phase during polycondensation with 1,4-phenylenediamine at the liquid-vapor interface.

The position of the molecular weight maximum of the curve may be expressed by q in Eq. (7.7):

$$q = \frac{C_{da}}{C_{ch}} \qquad (7.7)$$

C_{da} and C_{ch} are the concentrations in moles per liter of diamine and diacid chloride, respectively, which produce the maximum.

It was previously indicated [3] in cases of polycondensations in systems with two immiscible liquids that the value of q is invariable for a given system and independent of the level of reagent concentrations, and independent of which reagent concentration is varied. Thus, the ratio q may be calculated from the dependence of the specific viscosity of the polyamide on the initial concentration of diamine at a constant concentration of diacid chloride, or from the dependence of specific viscosity of the polymer on the concentration of diacid chloride at a constant concentration of diamine. The two values of q calculated in such a way practically coincide. Thus, q is a typical index for a given heterogeneous polycondensation system. The values of q for some cases of liquid-vapor interfacial polycondensation have been calculated (Table 7.4) [3]. The values obtained from different data sets, though, are in poorer agreement than has been found for liquid-liquid systems.

It is seen from Table 7.4 that the values of q for various heterogeneous systems are different. The value of q for a given liquid-vapor system, e.g., diamine (aqueous solution)-diacid chloride (vapor) is higher than in corresponding systems with two liquids.

7. Liquid-Vapor Interfacial Polycondensation

TABLE 7.4

Reagent Ratio Index, q, at Molecular Weight Maximum for Polyamide Condensations

System	q
Liquid-vapor systems (at 95°C)	
1,6-Hexamethylenediamine (aqueous solution)-oxalyl chloride (vapor)	50
1,6-Hexamethylenediamine (aqueous solution)-phosgene (vapor)	45-80
1,3-Phenylenediamine (aqueous solution)-oxalyl chloride (vapor)	14-29
1,4-Phenylenediamine (aqueous solution)-oxalyl chloride (vapor)	7-19
Liquid-liquid systems (at 20°C)	
1,6-Hexamethylenediamine (aqueous solution)-sebacoyl chloride (CCl_4)	5-6
Homogenous systems	
(Theoretical value)	1

Source: Ref. 3.

Data from Table 7.4 may be used to choose optimal conditions for polymer synthesis in liquid-vapor systems.

The concentration of a gaseous monomer in liquid-vapor interfacial polycondensations may be varied in two ways: (1) by a variation of monomer content at a constant absolute pressure of the gaseous mixture, or (2) by a change in the absolute pressure. It has been shown for instance, in the synthesis of 1,6-polyhexamethyleneoxamide at the liquid-vapor interface at 95°C, that the partial pressure of the gaseous reagent has a decisive influence on yield and intrinsic viscosity of the polymer [3]. The absolute pressure in a system affects the molecular weight and yield of a polymer only through the partial pressure in vapor phase.

V. FEATURES OF THE COURSE OF LIQUID-VAPOR INTERFACIAL POLYCONDENSATION

The accumulated data on polycondensation in liquid-vapor systems point to the dominant surface nature of the polycondensation processes. For highly reactive monomers in such systems reactions take place directly at the interface.

The above is evidenced by the character of the dependence of polymer molecular weight on temperature; by the dependence of the copolymer composition on the surface activity of the monomers; and by the sizable magnitude of the monomer ratio which produces the maximal polymer molecular weight.

It should be pointed out that in certain cases using one of the reagents in the form of gas or vapor does not guarantee that the polycondensation takes place at the liquid-vapor interface. Systems with a low polycondensation rate and with a gaseous reagent of high solubility in a reaction medium are likely to be liquid-liquid systems. Aromatic diamines and phosgene and systems with organic solvents give example of such systems, in which the solution of gaseous phosgene in a liquid phase takes place prior to polymer formation. Therefore, propagation of the macromolecules in this case occurs within the liquid volume rather than at the interface.

In order to choose successfully the optimal conditions of the synthesis of high-molecular-weight polymers in liquid-vapor systems it is essential to establish whether the process involved is a true liquid-vapor interfacial polycondensation or whether it proceeds within the liquid phase. For this aim, besides the above-mentioned dependence of the molecular weight on temperature, the site of reaction may be determined by analysis of the viscosity-yield correlation [8].

The possibility of using the dependences between molecular weight (viscosity) and yield of a polymer results from the following considerations. It is known [3] for polycondensation processes that average degree of polycondensation $\bar{n}$ is combined with conversion degree x by means of the equation:

7. Liquid-Vapor Interfacial Polycondensation

$$\bar{n} = \frac{1}{1-x} \tag{7.8}$$

This equation is true of the case when functional groups of monomers interact within a reaction volume in accordance with their chemical activity. It follows from this equation that, in the case of conversion degree equal to 50%, the main polycondensation products are dimers rather than high-molecular-weight polymers.

In the case of a process not occurring within this volume, as in heterogeneous systems, the dependence $\bar{n} = f(x)$ may differ from the dependence determined by Eq. (7.8) as a result of adsorption and diffusion factors. Specifically, in typical interfacial polycondensation it is possible to prepare a high-molecular-weight polymer at low degree of the reaction (x << 50%), and, further, it is possible to experience increase in molecular weight by decreasing the degree of the reaction.

Therefore, proceeding from the dependence character of $\bar{n} = f(x)$ a conclusion may be formed about the site at which the polycondensation process in heterogeneous systems is occurring. Using the viscosity (reduced, logarithmic, intrinsic) in place of molecular weight, and yield of polymer in place of conversion, the example below describes the procedure for plotting correlations which lead to qualitative conclusions on this question. The experimental dependences of the viscosity and yield of a polymer on any variable parameter (temperature, reagents concentration, amount of admixture, etc.) serve as the initial data for such correlations. The dependence of yield and intrinsic viscosity of poly(1,4-phenyleneoxamide) on the concentration of oxalyl chloride in a gaseous phase is given in Fig. 7.2a and 7.2b. The viscosity and yield data for six successive concentrations of oxalyl chloride are used for the construction of Fig. 7.2c. The above method for plotting and using the viscosity-yield correlation is described in detail in Ref. 8.

The polymer viscosity-yield correlations for polycondensations in various liquid-vapor systems are shown in Fig. 7.3a. (The straight-line average of data in Fig. 7.2c appears in Fig. 7.3a as curve 2.)

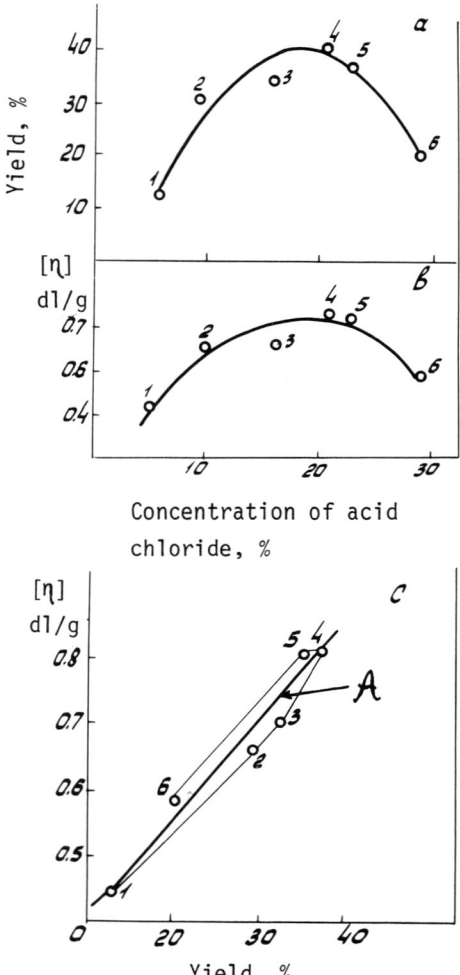

FIG. 7.2. Dependence of (a) yield of poly(1,4-phenyleneoxamide) and (b) intrinsic viscosity on oxalyl chloride concentration in vapor phase; (c) polymer viscosity-yield correlation in the liquid-vapor polycondensation of 1,4-phenylenediamine with oxalyl chloride at 95°C. (A: average curve.)

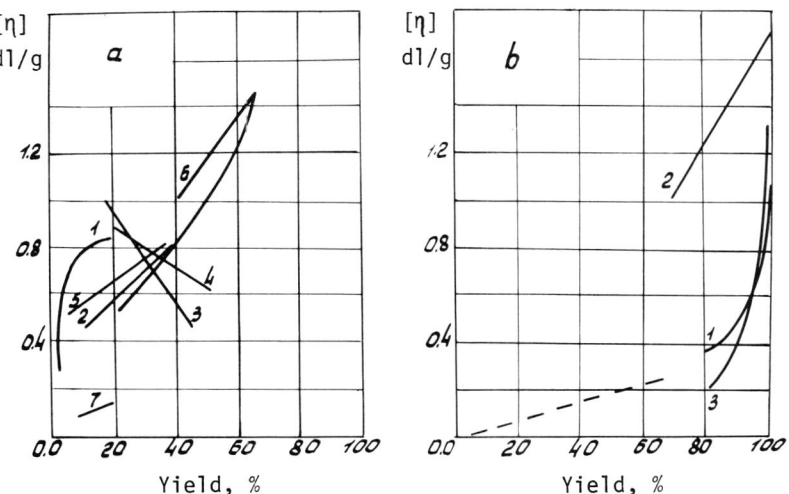

FIG. 7.3. Polymer intrinsic viscosity-yield correlations (a) for liquid-vapor systems and (b) for homogeneous and pseudohomogeneous liquid systems [8].
  a. 1: 1,6-Hexamethylenediamine (water)-oxalyl chloride (vapor); a variable parameter (v.p.), reaction temperature. 2: 1,4-Phenylenediamine (water)-oxalyl chloride (vapor); v.p., oxalyl chloride concentration in vapor phase. 3: 1,6-Hexamethylenediamine (water)-oxalyl chloride (vapor); v.p., diamine concentration in aqueous phase. 4: 1,6-Hexamethylenediamine (water)-oxalyl chloride (vapor); v.p., oxalyl chloride concentration in gaseous phase. 5: 1,6-Hexamethylenediamine (water)-phosgene (vapor); v.p., diamine concentration in aqueous phase. 6: 1,4-Phenylenediamine (water)-oxalyl chloride (vapor); v.p., diamine concentration in aqueous phase. 7: 1,4-Phenylenediamine (water)-phosgene (vapor); v.p., reaction temperature.
  b. 1: Ethylene glycol-4,4'-diphenylmethane diisocyanate (homogeneously); v.p., solvent in synthesis. 2: 2,5-dimethylpiperazine-sebacoyl chloride (homogeneously in chloroform); v.p., acceptor HCl. 3: m-Phenylenediamine (water)-isophthaloyl chloride (tetrahydrofuran); v.p., monomer ratio.
  Yield is based on the amount of diacid chloride.

In Fig. 7.3b, the data from polycondensations in two homogeneous systems are included for comparison. Also included are the data from a liquid-liquid polycondensation in which it was confirmed by special tests that the polycondensation occurred within a liquid phase and not as an interface.

Comparison of Fig. 7.3a and 7.3b shows that the viscosity-yield relationships for liquid-vapor systems differ sharply from homogeneous and heterogeneous liquid systems where the polycondensation process takes place within a liquid phase. In liquid-vapor systems, high-molecular polymer is obtained in yield less than 50%, whereas in homogeneous and pseudohomogeneous systems (Fig. 7.3b), high-viscosity polymer is obtained only at high yields, e.g., >80%. Also, there are cases of anomalous correlations in liquid-vapor systems whereby a decrease in the viscosity of the polymer is observed with increase in its yield (curves 3 and 4 in Fig. 7.3a). It can be seen that such inverse patterns are observed for the very reactive reactant pairs. Such anomalous cases are not observed in polycondensations proceeding within liquid phases. Fig. 7.3a indicates that polycondensation in liquid-vapor systems does not occur within the liquid phase. Therefore, it can be said that polycondensations in liquid-vapor systems proceeds very close to the interface in agreement with the diffusion-absorption mechanism.

Although a straight-line average (curve A in Fig. 7.2c) reasonably represents all the data and successfully combines the data on either side of the maximum point, the two legs of curve 6 in Fig. 7.3a about the maximum data are more distinctly separate. This separation is caused by change in the kinetics of the process, following from variation in the concentrations of 1,4-phenylenediamine in reaction medium. It is evident that additional variations in the ratio between factors (diffusion, adsorption, kinetic, hydrolytic) composing the process occur as result of change in the concentration of 1,4-phenylenediamine.

Curve 7 of Fig. 7.3a is distinct. It gives the results for the system of 1,4-phenylenediamine (water)-phosgene (vapor). The

7. Liquid-Vapor Interfacial Polycondensation

viscosity-yield correlation for this system differs from those of the other liquid-vapor systems and is analogous to the correlation of homogeneous systems. It suggests that the polycondensation occurs within the aqueous phase. Its temperature dependence of polymer viscosity, the low rate of the main reaction, and the high solubility of phosgene in water also indicate that, in this case, the polycondensation proceeds within the liquid phase rather than at an interface.

The viscosity-yield correlation of the polycondensation intimates the location of the reaction zone qualitatively (the volume of a liquid phase vs. the interface).

For liquid-vapor interfacial polycondensation it is rather interesting and important to know why aqueous solutions at high temperatures present optimum conditions for the synthesis of high-molecular polymers. These conditions would appear to predicate the complete hydrolysis of the diacid chlorides used in the process.

Successful interfacial polycondensation occurs with aqueous solution-vapor systems at elevated temperatures for the following reasons: (1) there is a pronounced acceleration of the propagation reaction in the aqueous media; (2) a significant value of the surface tension exists at the interface; and (3) there is a suppression of polycondensations proceeding within the liquid phase by intensified hydrolysis.

Let us consider each of the reasons in detail.

The significant acceleration of the acylation reaction of amines with acid halides in the presence of water has recently been established by direct experiments. Table 7.5 shows the influence on the reaction rate constants of isophthaloyl difluoride with aniline (k_a) and with water (k_h) of the composition of water-tetrahydrofuran-solvent mixtures [9].

The rate constant of the acylation reaction of aniline in an aqueous-organic mixture are hundreds to thousands of times larger than for the reaction proceeding in organic solvents. The large acceleration of the acylation reaction has been established for a series of amines and acid halides in various systems of organic

TABLE 7.5

Dependence of Rate Constants of Isophthaloyl
Difluoride with Aniline (k_a) and with Water (k_h)
Content in Mixtures of Tetrahydrofuran-Water at 25°C

Water content (vol %)	$k_a \times 10^4$ (liter/mol sec)	$k_h \times 10^4$ (liter/mol sec)	$\dfrac{k_a}{k_h}$
0	13	—	—
6	38	0.17	220
10	76	0.31	250
25	310	0.47	660
40	700	0.63	1110
50	1150	0.73	1570

Source: Ref. 9.

liquid-water [10-11], besides the case given in Table 7.5. The form of the dependence of acylation rates of amines by acid halides on water content in reaction media is shown in Figure 7.4. This dependence is observed for the following cases of acylation: 4-anisidine + 4-toluenesulfonylchloride in medium of tetrahydrofuran-water [10]; 4-anisidine + 4-toluenesulfonylchloride in medium of acetone-cyclohexanone-water [10]; aniline + isophthaloyl fluoride in medium of tetrahydrofuran-water [11]; and in others [11]. If the data from Refs. 9-11 are extrapolated to 100% water content, the difference between the acylation reaction in water and that in organic solvents such as tetrahydrofuran will be about a thousandfold. It should be noted that water accelerates the hydrolysis of acid halides to a lesser degree than it accelerates the acylation reaction of amines (Table 7.5).

From a study of the effect of medium on the formation rate of amides of carboxylic and arylsulfonic acids [12] one may assume a mechanism for the accelerating action of water in the acylation reaction involving structures of activated complexes and water:

7. Liquid-Vapor Interfacial Polycondensation

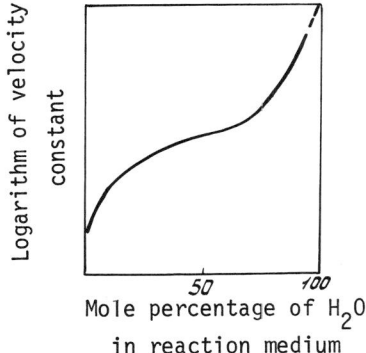

FIG. 7.4. Dependence of acylation rate of aromatic amines by acid halides on water content (according to data in Refs. 9-11) in water-organic liquid media. Additional definition in text.

```
           a                              b
          R   Ar                         R   Ar
          |   |  δ+                      |   |  δ+
    δ-    |   |                    δ-    |   |
      O = C---N---H                   O = C---N---H
          |   |                         |   |
          X   H                         X   H
          |   |                         |   |
         HOH  O---H                    HOH---O---H
              |                              |
              H                              H
```

The above HX cleavage is facilitated by water resulting in increasing the propagation rate of the polymer chain and, therefore, in increasing the polymer molecular weight. At the same time the complex formation tripart may explain why acceleration of the acylation process occurs in water to a greater extent than the acceleration of hydrolysis.

The value of the interfacial surface tension is the second important factor responsible for the success of liquid-vapor interfacial systems. According to Ref. 13 the value of the surface tension must be not less than 10 dyne/cm to permit successful liquid-liquid interfacial polycondensation. Practically, all liquid-vapor systems meet this requirement; for instance, the surface tension at the interface between water and nitrogen is equal to 72 dyne/cm.

One of the basic functions of the existence of an interface in heterogeneous polycondensations is to produce monomer countercurrents and increase the interfacial monomer concentration [14]. Because of such countercurrents it is possible to form high-molecular-weight polymer in non-equimolar monomer relations. A high surface-tension value is the basic factor ensuring the direction of flow of the reactive compounds in dynamic systems, e.g., those undergoing bubbling or foaming.

Now, let us consider the third important factor in liquid-vapor interfacial polycondensations—the suppression of polymer formation within a liquid phase.

In the case of a liquid-vapor system the process within the liquid phase produces a low-molecular-weight product. A high-molecular-weight product would require an equimolar stoichiometry between the reagents to within at least 0.5% by mole. It is practically impossible to achieve this for the whole volume of a liquid phase in liquid-vapor systems. Therefore, the same factors that suppress the reaction within the liquid phase will promote a relative share of high-molecular-weight products in the liquid-vapor system.

The probability of a reaction occurring outside a water phase in the case of liquid-vapor systems is less than that in two-phase liquid systems due to absence of an organic phase. The concept of polycondensation occurring in the gaseous phase is also not favored.

The suppression of the reaction within the aqueous phase is assisted by the high temperature of the process (e.g., up to 95°C) because of the intensified hydrolytic reaction of the highly reactive diacid halide. This idea is supported by data in Table 7.1, from which it is seen that the molecular weight of polymers in liquid-vapor systems is higher than in the synthesis of the same polymer utilizing a two-liquid interfacial system if the monomer used is readily hydrolyzable.

The data on polycondensation of diamines with oxalyl chloride in organic liquid-vapor systems (Fig. 7.5) are of interest from the above stated point of view. It is seen from the figure that polycondensation in systems of diamine (organic liquid)-oxalyl chloride (vapor) is

7. Liquid-Vapor Interfacial Polycondensation

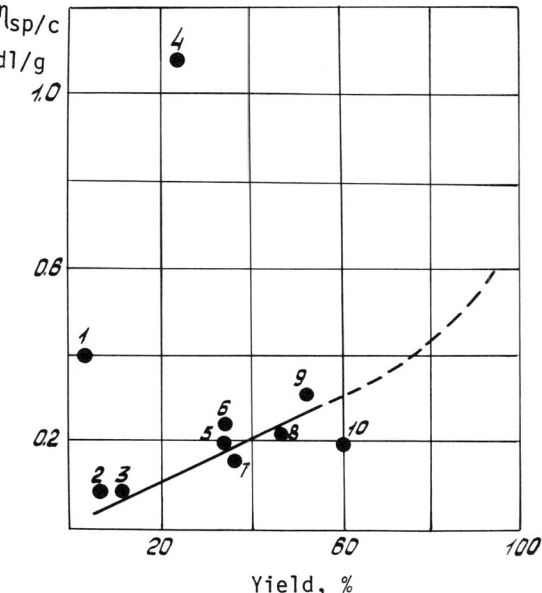

FIG. 7.5. Polymer inherent viscosity-yield correlation in the polycondensation of 1,6-hexamethylenediamine (liquid) and oxalyl chloride (vapor) for various liquid-vapor systems. Liquid phase: 1, dimethylformamide; 2, ethanol; 3, n-butanol; 4, water; 5, dibutyl ether; 6, xylene; 7, dioxane; 8, nitrobenzene; 9, n-octane; 10, chlorobenzene.

rather like polycondensations occurring within a liquid phase. This is actually the case owing to the high solubility and absence of intensive hydrolysis of oxalyl chloride in organic liquids.

Systems employing as the liquid phase water or dimethylformamide, i.e., liquids that activate to oxalyl chloride, are the most outstanding among the organic liquid-vapor systems: in these liquids the formed polymer has higher molecular weight than in inert organic solvents. Therefore, water produces hydrolysis of acid chlorides, hinders the formation of low-molecular-weight polymer within the volume of liquid phase, allows high-molecular-weight polymer formation at the interface, but lowers the yield of the interfacial polymer.

VI. CONCLUSION

As in other variants of interfacial polycondensation liquid-vapor polycondensation has an important technological feature: it does not require very exact metering of monomers in the synthesis of high-molecular-weight polymers.

Liquid-vapor interfacial polycondensations differ in their simplicity from liquid-liquid interfacial systems. In the former system only one component—water—is usually present besides the monomers. This method also permits the use in polycondensation of readily hydrolyzing monomers, which is practically impossible in other aqueous systems.

The auxiliary procedures in liquid-vapor interfacial polycondensation are simplified considerably. In this case, only ionic impurities (i.e., NaCl) needs to be washed from the polymer product, while in other polycondensation processes the separation procedure of an organic solvent from the polymer can be rather involved (steam distillation, etc.). Solvent recovery is not required in the synthesis of polymers by the liquid-vapor interfacial polycondensation. Recovery or recirculation of diamine from the aqueous phase may present some technical difficulties in liquid-vapor interfacial polycondensations.

The polymers obtained by liquid-vapor interfacial polycondensation contain strongly adsorbed water which should be removed prior to further polymer characterization or processing. This can be achieved by means of effective drying, for example, under vacuum.

Besides the technique of introducing one reagent into the reaction by means of bubbling, a foaming process may be used for liquid-vapor interfacial polycondensation [15]. For this technique, polymer may be synthesized in a foam generator. Foam generators are efficient vessels, suitable to large-scale manufacturing.

REFERENCES

1. L. B. Sokolov, L. V. Turetskii, and T. V. Kudim, *Vysokomol. Soedin.*, 2:1744 (1960).
2. L. B. Sokolov, *J. Polymer Sci.*, 58:1253 (1962).
3. L. B. Sokolov, *Synthesis of Polymers by Polycondensation*, Israel Program for Scientific Translation, Jerusalem, 1968.
4. P. W. Morgan, *Polycondensation Polymers by Interfacial and Low Temperature Methods*, Interscience, New York, 1966.
5. L. B. Sokolov and L. V. Turetskii, *Vysokomol. Soedin.*, 7:1997 (1965).
6. V. Z. Nikonov and L. B. Sokolov, *J. Phys. Chem.* (Russ.), 43:1039 (1969).
6a. V. V. Korshak and S. V. Vinogradova, *Irreversible Polycondensation* (Russ.), Nauka, Moscow, 1972.
7. N. K. Vorobyev and L. B. Kuritsyn, *Izv. Vyssh. Ucheb. Zaved. (Communications of the Institutions of Higher Education)*, 6:591 (1963).
8. L. B. Sokolov, *Vysokomol. Soedin., Ser. A,* 12:971 (1970).
9. L. B. Sokolov, L. I. Litvinenko, V. I. Logunova, S. S. Medved, V. I. Savelova, and N. M. Chentsova, *Vysokomol. Soedin.*, 13:357 (1971).
10. V. A. Savelova, L. M. Litvinenko, N. M. Chentsova, L. B. Sokolov, S. S. Medved, A. P. Popov, and V. I. Tokarev, *Reactivity Org. Compounds* (Russ.) 7:847 (1970).
11. L. B. Sokolov, V. I. Logunova, D. F. Sokolova, *Dokl. Akad. Nauk SSSR*, 189:343 (1969).
12. L. M. Litvinenko and V. A. Savelova, *Reactivity Org. Compounds* (Russ.), 5:838, 862 (1968).
13. N. V. Mikhailov, V. I. Maiboroda and S. S. Nikolayeva, *Vysokomol. Soedin.*, 2:991 (1960).
14. Yu. V. Sharikov, M. I. Fedotova, and L. B. Sokolov, *Vysokomol. Soedin.*, 15:982 (1973).
15. I. V. Parfenov, L. B. Sokolov, and S. S. Novokresshchenov, *J. Appl. Chem.* (Russ.), 39:208 (1966).

8.

COPOLYCONDENSATION AND MACROSCOPIC KINETICS

L. B. Sokolov
Department of Polycondensation
Processes, All-Union Research
Institute of Synthetic Resins
Vladimir, USSR

V. Z. Nikonov
Department of Polycondensation
Processes, All-Union Research
Institute of Synthetic Resins
Vladimir, USSR

I.		INTRODUCTION	168
II.		FOUNDATIONS OF MACROSCOPIC KINETICS FOR COPOLYCONDENSATION IN TWO-PHASE SYSTEMS	170
III.		COPOLYMER COMPOSITION IN DIFFERENT MACROSCOPIC REGIONS	175
	A.	Copolymer Composition in Kinetic Region of Copolycondensation	175
	B.	Copolymer Composition in Copolycondensation Limited by Mass Transfer Processes	181
	C.	Copolymer Composition in Transitional Copolycondensation Regions	190
	D.	Copolymer Composition under Other Possible Copolycondensation Circumstances	192
IV.		EFFECTS OF SIDE REACTIONS ON COPOLYMER COMPOSITION	193
V.		NONUNIFORMITY OF COPOLYMER COMPOSITION UNDER DIFFERENT VARIANTS OF MACROSCOPIC KINETICS	196
VI.		PRACTICAL METHODS AND THE CONDITIONS OF VARIOUS COPOLYMER SYNTHESES	197
		REFERENCES	200

I. INTRODUCTION

Copolycondensation is an important method for the production of new polymers. The properties of polymer materials can be widely varied by means of copolycondensation. However, the theoretical and, in particular, the quantitative foundations of copolycondensation processes have not been adequately developed. The lack of development of a theory of polymer synthesis by polycondensation is in sharp contrast to the situation of addition polymerization where theoretical aspects have been developed more fully [1]. Further development of nonequilibrum copolycondensation theory has recently been made, and the basic results of the research work on solution copolycondensation are given in the review by Vasnev and Kuchanov [1a] and in some other papers [1b, 1c, 1d].

The important question of the theoretical basis of polycondensation in relation to kinetic processes in two-phase liquid systems will be discussed briefly.

The simplest case of copolycondensation is one in which two monomers with the same functional groups (comonomers) participate in the macromolecule formation. In this case the accretion of a polymer chain occurs as a result of two propagation reactions with different rates. This simplest case, called interbipolycondensation [2], may be represented as follows:

$$A + B + C \longrightarrow ACBCBCA\ldots \qquad (a)$$

where A and B are the bifunctional compounds with the same functional groups (comonomers; for example, two diamines) and C is the "linking" reagent having functional groups of another type (intermonomers; for example, diacid chloride). In polycondensation, following the scheme (a), a "duplex" monomer residue from copolycondensing monomers and intermonomer is the repeating unit of a chain.

In addition to case (a) there are also the other variants of polycondensation processes; for example, bipolycondensation (for instance, two amino acids):

$$A + B \longrightarrow AABABB\ldots \qquad (b)$$

8. Copolycondensation and Macroscopic Kinetics

and multipolycondensation (for instance, several amino acides).

$$A + B \cdots + P \longrightarrow A\cdots A\cdots BPPB\cdots AP \qquad (c)$$

More complex cases of copolycondensation may be observed in which one monomer takes part both in polycondensation and in other reactions, such as polyamidation reactions with simultaneous hydrolysis of acid dihalides, the polycondensation being followed by the formation of cyclic products, etc.

Polycondensation copolymers are normally characterized by chemical composition, molecular weight and its distribution, regularity of the distribution of units over a chain, and composition nonuniformity. The chemical composition of a copolymer is usually expressed in molar fractions (percentages) of one or another unit, and, for example, is determined on the basis of elementary analysis.

One should differentiate between copolymers with alternate unit distribution, random, and block-copolymers, in terms of regularity of distribution of units over a chain of linear copolymers. Composition nonuniformity means that copolymer macromolecules differ in composition.

The ratio of rates of inclusion of the monomers into a copolymeric chain, defined by kinetic or other factors, is the major factor in describing copolycondensation. In the quantitative approach to copolycondensation, some simplifications, which do not change the physical picture of the processes, are usually adopted due to the complexity of the kinetic equations. Firstly, the principle involved is that the endgroup activity is independent of the length of a macromolecule to which it is joined (Flory's principle). In this paper both groups of bifunctional monomers are assumed to have the same reactivity [2]. Secondly, it is assumed that the propagation stage is of a bifunctional nature, and kinetic equations of the second order are thus applicable.

II. FOUNDATIONS OF MACROSCOPIC KINETICS FOR COPOLYCONDENSATION IN TWO-PHASE SYSTEMS

Macroscopic kinetics investigates the processes of chemical transformation occurring together with processes of mass and heat transfer in heterogenous systems [3].

The "process" of macroscopic kinetics refers to a certain combination of two main characteristics: the physical region in which the chemical reaction occurs (e.g., volume or the surface of an interface), and the nature of the process which determines the total reaction rate (e.g., diffusion or chemical kinetics).

Depending on the combination of the above characteristics the following process types of macroscopic kinetics may be differentiated (following D. A. Frank-Kamennetsky, Ref. 3) in two-phase liquid systems:

1. The *internal kinetic* process: chemical reaction occurs in the bulk reaction phase and is limited by the chemical reaction rate.
2. The *external kinetic* process: chemical reaction occurs at an interface, containing equilibrium concentrations of reagents, and is limited by the chemical reaction rate.
3. The *internal diffusion* process: chemical reaction occurs in a zone of the bulk phase and is limited by diffusion of reagents within the reaction phase.
4. The *external diffusion* process: chemical reaction occurs at an interface, containing nonequilibrium concentrations of reagents, and is limited by diffusion of reagent from the unreactive bulk phase to the interface.

The type of process occurring is determined by the ratio between chemical kinetics and diffusion factors. These process types should also be observed in interfacial polycondensation. The *process stage* is an unseparated physical or chemical phenomenon caused by the regular change of the process rate.

The term *limiting stage of the process* is of great importance. It is understood to be the stage that determines the total rate of the process.

8. *Copolycondensation and Macroscopic Kinetics*

The major problem of macrokinetics involves finding the expression for the overall process rate and solving the combined equations for diffusion and chemical interaction.

In interbipolycondensation occurring in a two-phase system an equation may be written for each comonomer which combines the rate of inclusion of monomer into a polymeric chain and its initial concentration in the phase. The concentration in the initial phase, A, is a certain function, f(i), of two-phase system parameters, i, which is independent of monomer concentration and is a constant for given process, may be expressed in such an equation as a product. For instance, for monomer A:

$$-\frac{d[A]}{dt} = f(i)\,[A] \tag{8.1}$$

The relative rates of inclusion of monomers into a polymeric chain are defined by the equation

$$\frac{d[a]}{d[b]} = r\,\frac{[A_o] - a}{[B_o] - b} \tag{8.2}$$

in which $[A_o]$, $[B_o]$ are the initial concentrations of comonomers with the same type of functional groups in the system (or in the initial phase); $[a]$, $[b]$ are the quantities of monomers incorporated into the copolymer, expressed as concentrations relative to the total volume of the system (or in the initial phase); and r is the copolycondensation parameter, here equal to a function $f^A(i)/f^B(i)$.

The integral form of the Eq. (8.2) is as follows:

$$\frac{[A_o] - [a]}{[A_o]} = \left\{\frac{[B_o] - [b]}{[B_o]}\right\}^r \tag{8.3}$$

or

$$r = \frac{\log([A_o] - [a]) - \log[A_o]}{\log([B_o] - [b]) - \log[B_o]} \tag{8.4}$$

The differential form of Eq. (8.2) may be used for calculation of the relative rate of inclusion of monomers into the polymer chain only at low degrees of conversion (not more than 5% of the copolymerizable monomers). Equations (8.3) and (8.4), i.e., the integral equations for copolymer composition, may be used at degrees of conversion up to 50-90%. At >90% degree of copolymer conversion the value of r obtained from experimental data is less accurate.

Calculation of r from experimental data is carried out as follows. First, the amounts a and b of monomers that have reacted are determined from the yield and composition of the copolymer being studied, for example, from elementary analysis. Then r is calculated utilizing the known values of $[A_o]$ and $[B_o]$ from Eq. (8.4).

Table 8.1 shows experimental values of the copolycondensation parameter r for copolyamide synthesis in a number of systems. Values of r for the interfacial synthesis of polycarbonates are reported elsewhere [4-9]. Systematization of these data is hampered due to the high degrees of conversion and the errors resulting from this.

During investigation of interfacial polyamidation the copolycondensation parameter r for some monomer pairs was shown to depend on the conditions of synthesis [14].

The dependence of r on stirring intensity is of great interest in connection with discussions of the kinetics of copolycondensation reactions. The copolymer composition and copolycondensation parameter change [13, 15] as one goes through the transition from a copolycondensation process without stirring to one with stirring, as shown in Table 8.2 [13]. The dependence of r on stirring rate is clear evidence of the mass transfer effect (including diffusion factors) on the copolycondensation process in two-phase systems. The macroscopic kinetics of copolycondensation processes in two-phase systems is very complex. Therefore, the quantitative study of copolycondensation in two-phase systems is possible only for certain cases utilizing simplifying assumptions. The (practically) absolute irreversibility of interfacial copolycondensations is one of these assumptions.

It should be noted that the competition between diffusion and chemical factors must be taken into account in studying irreversible copolycondensation processes. Thus, in obtaining high-molecular-weight polymers by reversible polycondensation processes, wherein

TABLE 8.1

Copolycondensation Parameter r of Interphasic Synthesis of Copolyamides in various Organic Solvent-Water Systems

Monomer A	Monomer B	Intermonomer	Organic solvent	Hydrodynamics	r	Ref.
Ethylenediamine	1,6-Hexamethylene-diamine	Terephthaloyl chloride	Toluene	Stirring	0.48	10
	1,3-Phenylenediamine	Adipoyl chloride	Benzene		0.17	11
			Benzene		0.41	11
Tetrachlorterephthaloyl chloride	Terephthaloyl chloride	1,6-Hexamethyl-enediamine	Toluene		0.21	10
Adipoyl chloride	4,6-Dibromisophthaloyl chloride		Benzene	Without stirring	0.53	12
Sebacoyl chloride					0.77	12
Isophthaloyl chloride					1.05	12
Terephthaloyl chloride					1.17	12
Adipoyl chloride	Isophthaloyl chloride			Stirring	0.66	12
	4,6-Dibromisophthaloyl chloride				0.66	12
Terephthaloyl chloride		1,3-Phenylene-diamine			1.20	12
Adipoyl chloride					0.94	12
		Piperazine			1.4	12

TABLE 8.2

Effect of Stirring Intensity on Copolymer
Composition and Copolycondensation Parameter r in the System of
Tetramethylenediamine(A)-Decamethylenediamine(B)-Isophthaloyl Chloride

Hydrodynamic conditions	Molar fraction A		η_{sp} in H_2SO_4 0.5 g/dl	$r \pm \Delta r$
	In initial mixture of diamines	In copolymer		
Without stirring	0.75	0.15	0.57	0.04 ± 0.02
	0.75	0.39	0.64	0.18 ± 0.05
	0.75	0.33	0.77	0.13 ± 0.05
				$r_{av} = 0.12$[a]
Slow stirring (300 rpm)	0.75	0.51	0.51	0.29 ± 0.08
	0.75	0.29	0.65	0.08 ± 0.03
	0.75	0.19	0.55	0.04 ± 0.02
				$r_{av} = 0.14$[a]
Intensive stirring (5,000 rpm)	0.75	0.68	0.62	0.58 ± 0.22
	0.75	0.68	0.57	0.57 ± 0.22
	0.75	0.74	0.62	0.90 ± 0.36
				$r_{av} = 0.68$[a]

[a]Average values of r are indicated for parallel tests.

maximal molecular weight is determined by the value of the polycondensation equilibrium constant, the diffusion factors seldom determine the total rate of the process. The supposition about the independence of diffusion currents of monomers is the second important assumption. This assumption has been experimentally confirmed for the diffusion of monofunctional compounds at moderate concentrations [16].

The theoretical basis of copolycondensation under different variants of macroscopic kinetics will be discussed in the following sections.

8. Copolycondensation and Macroscopic Kinetics

III. COPOLYMER COMPOSITION IN DIFFERENT MACROSCOPIC REGIONS

It is becoming apparent, as noted above, that the macroscopic region of a process must decidedly affect the copolymer composition in the interfacial polycondensation.

A. Copolymer Composition in Kinetic Region of Copolycondensation

In the case of relatively slow propagation reactions, the copolymer composition is related to the composition of the initial mixture by the kinetic characteristics of the reagents, i.e., it is determined by the monomer reactivity. In copolycondensation, as in vinyl copolymerization, the copolymer composition is defined by relative rates of monomer addition to a polymer chain, i.e., by the competition of the propagation reactions of various reagents.

In the case of copolycondensation of two similar monomers, A and B (e.g., two diamines), with the intermonomer, C (e.g., an acid dihalide), the rates of incorporation of these monomers into the condensation copolymer can be expressed by the following equations:

$$-\frac{d[A]}{dt} = k_1[A][C] \tag{8.5}$$

$$-\frac{d[B]}{dt} = k_2[B][C] \tag{8.6}$$

where t is the time and k_1 and k_2 are the constants of the respective propagation reactions.

The relationships between reaction rates (from Eqs. 8.5 and 8.6), and consequently the ratio of molar monomer concentrations in copolymer a and b will be:

$$\frac{a}{b} \simeq \frac{d[a]}{d[b]} = r_o \frac{[A]}{[B]} \tag{8.7}$$

where $r_o = k_1/k_2$. Equation (8.7) is the differential equation describing the polycondensation product composition in the kinetic region.

If the copolymer composition is expressed in terms of initial monomer concentrations $[A_o]$ and $[B_o]$ and the resultant equation integrated, we obtain:

$$\frac{[A_o] - [a]}{[A_o]} = \left\{ \frac{[B_o] - [b]}{[B_o]} \right\}^{r_o} \tag{8.8}$$

The equation was derived independently by Streichman and by Beste [17, 18]. Analysis of Eq. (8.8) [10, 17-19] has indicated that it is hardly suitable for calculating r_o values of a kinetic region. Really, formation of high-molecular-weight copolymer is possible only at high degrees of conversion, and in this case the copolymer composition will be equal to the initial mixture composition, while the copolycondensation at low degrees of conversion will result in low-molecular-weight product.

Equations (8.7) and (8.8) do not take into account the heterogeneity of the process and its macroscopic region and are applicable only to irreversible polycondensations which proceed homogeneously in solution.

1. Internal Kinetic Process

The most common arrangement of copolycondensations proceeding in heterogeneous systems in an internal kinetic process is one wherein monomers A and B, initially dissolved in phase I, diffuse into phase II, which contains an intermonomer C, insoluble in phase I (Fig. 8.1). If the partition equilibrium of monomers in each phase is reached, i.e., the rate of partition equilibrium in a system is much higher than the rate of chemical reaction, it is evident that, in addition to the terms in Eq. (8.7), the distribution coefficients of monomers between phases and phase volume must be included in an equation for the copolymer composition:

$$\frac{d[a]}{d[b]} = r_o \frac{(1/K_D^B + V''/V')\{[A_o] - [a]\}}{(1/K_D^A + V''/V')\{[B_o] - [b]\}} \tag{8.9}$$

Besides known values a, b, r_o, A_o, B_o, this equation includes K_D^A and K_D^B which are the distribution coefficients of monomers between

8. Copolycondensation and Macroscopic Kinetics 177

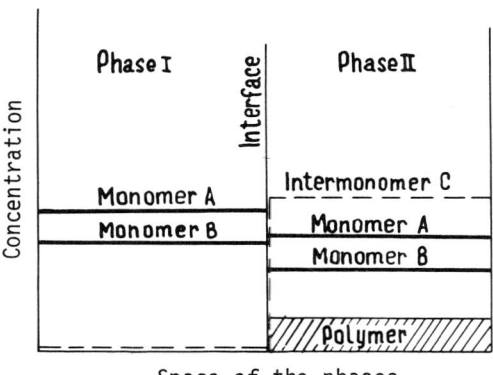

FIG. 8.1. Diagram of the internal kinetic process of copolycondensation. Jump in concentrations at the interface corresponds to distribution coefficients; $K_D < 1$.

phases II and I for monomers A and B, respectively, and V', V" are the volumes of the phases.

Equation (8.9) holds for relatively slow propagation reactions proceeding in two-phase systems, as well as for copolycondensation of highly reactive monomers at very low concentrations. Under such conditions the molecular weight of a copolymer is strong affected by the monomer ratio [17-19] and therefore, because of excess or paucity of the monomers, oligomeric products are obtained. In this case the high-molecular-weight copolymer composition is equal to that of the initial mixture.

In the case where K_D^A and $K_D^B \rightarrow \infty$, i.e., the reaction completely proceeds in phase II while in phase I the monomers are absent, Eq. (8.9) will transform into Eq. (8.7) for the homogeneous process. This is a case of so-called emulsion copolycondensation, i.e., heterogeneous systems wherein the polymer formation occurs in only one of the reaction phases. Examples of this are the polycondensation in m-phenylenediamine with aromatic carboxylic diacid chlorides in systems of tetrahydrofuran-water-Na_2CO_3 or in cyclohexanone-water [20, 21] and also the polycondensations involving water-insoluble diamines.

TABLE 8.3

Distribution Coefficients of
Diamines in Some Two-Phase Systems

Diamine	System	K_D
Aliphatic diamines		
1,4-Tetramethylenediamine	Chloroform-water	0.05
1,10-Decamethylenediamine		13.0[a]
Aromatic diamines		
m-Phenylenediamine	Tetrahydrofuran-(water + Na_2CO_3)	8-10
	Tetrahydrofuran-(water + NaCl)	4-6
	Propylene oxide-(water + Na_2CO_3)	4.9
	Cyclohexanone-water	5.8
4,4'-Diamino-3,3'-dimethyldiphenyl		70
	Benzonitrile-water	180

[a] K_D of decamethylenediamine depends on the concentration.

The distribution coefficients of diamines for some systems are listed in Table 8.3. In the case of copolycondensation of these ("emulsion") systems the composition of the high-molecular-weight copolymer is equal to that of the initial mixture

The kinetic process may be achieved practically for all interfacial copolycondensations by very rapid stirring, while reducing the concentration of reagents in the reaction phase (i.e., intermonomer concentration) will have the result that the rate of monomer transfer from the nonreaction phase should exceed considerably the reaction rate.

Quantitative data for the above indicated conditions of interfacial copolycondensation in the kinetic process have been obtained [14]. Concerning the copolycondensation of 1,4-tetramethylenediamine

8. Copolycondensation and Macroscopic Kinetics

(A) and 1,10-decamethylenediamine (B) (intermonomer-isophthaloyl chloride) in the chloroform-water system, the experimental value of r for this system was found to be 0.03.

According to Eq. (8.9), for the case of copolycondensation in a kinetic process, a theoretical value of the copolycondensation parameter is:

$$r' = r_o \frac{(1/K_D^B + v''/v')}{(1/K_D^A + v''/v')} \qquad (8.10)$$

It is necessary for the calculation of r' to know K_D^A, K_D^B, and r_o (or, as is shown in Ref. 2, the relative reactivity of the primary groups of monomers). The relative reactivity r_o, of the primary groups of tetramethylenediamine and decamethylenediamine have been determined by special experiments [13]. In these experiments, the reaction of diamines with isophthaloyl chloride (or isophthaloyl fluoride) was carried out at -60° to -70°C in chloroform while stirring was vigorous to create conditions where the rate of the process was determined by the kinetics of chemical reaction rather than by the mixing rate. Calculation of r_o was performed using Eq. (8.7) with separate determinations of the initial and final contents of free diamines and their hydrochlorides. The relative reactivity of the primary groups of tetramethylene toward decamethylene was ~1.9. Distribution coefficients for these diamines, determined by titration of both water and organic phases, are given in Table 8.3 (for decamethylene, K_D depends on concentration, an intermediate value being taken for calculations). The value of r = 0.029 was obtained as a result of the substitution in Eq. (8.10) of numerical values of r_o, distribution coefficients, and phase volumes, depending on the conditions of polymer synthesis. The coincidence of the calculation value of r' with the experimental value of the copolycondensation parameter (r = 0.03) is satisfactory. This coincidence reveals a kinetic nature of the copolycondensation of tetramethylenediamine and decamethylenediamine in a chloroform-water system under the follwoing experimental conditions: strong stirring (n = 5,000 rpm), an overall initial concentration of diamines in the water phase of 0.02 mol/liter, and an initial concentration of isophthaloyl chloride in chloroform of 0.02 mol/liter.

2. External Kinetic Process

Interfacial copolycondensation in the external kinetic process, i.e., when the reaction proceeds at the interface and is determined by the reaction rate on the surface, appears to be theoretically possible. In this case $K_D^A \approx 0$ and $K_D^B \approx 0$, i.e., the monomers are completely insoluble in phase II. Here the equation of copolymer composition is as follows:

$$\frac{d[a]}{d[b]} = r' \frac{[A_0] - [a]}{[B_0] - [b]} \qquad (8.11)$$

The following is an explanation of the sense of r'. Equation (8.12) can be derived from the law of active surfaces [22, 23] for Eq. (8.11):

$$r' = r_0^S \frac{b_A \cdot n_S^A}{b_B \cdot n_S^B} \qquad (8.12)$$

where r_0^S is the relative reactivity constant of monomers on the surface, b_A, b_B are the adsorption coefficients, and n_S^A, n_S^B are the numbers of adsorption sites per unit surface area for monomers A and B, respectively. The adsorption coefficients b_A, b_B serve as a distribution factor between the reactive surface and the bulk since the ratio of the surface monomer concentrations $[A]'/[B]'$ is

$$\frac{[A]'}{[B]'} = \frac{n_S^A \theta_A}{n_S^B \theta_B} = \frac{n_S^A b_A [A]}{n_S^B b_B [B]} \approx \frac{\Gamma_A}{\Gamma_B} = \frac{n_S^A \theta_A}{n_S^B \theta_B} \qquad (8.13)$$

where θ_A, θ_B are the parts of interface area occupied by monomers A and B, respectively, and Γ_A, Γ_B are the quantities of monomers A and B, respectively, per unit interface area.

For experimental determination of the above parameters see Sect. III.B.3.

The external kinetic process of interfacial copolycondensation, i.e., limiting the process by the rate of the chemical reaction

8. *Copolycondensation and Macroscopic Kinetics* 181

occurring at the interface, has not yet been experimentally detected.
Copolycondensations proceeding in this process may be assumed, for
example, in the process where two monomers adsorbed on the surface
interact with an intermonomer, slowly diffusing to the interface.

B. *Copolymer Composition in Copolycondensation Limited by Mass Transfer Processes*

It was previously shown (see Sect. II) that the mass transfer process
can be a limiting factor of copolycondensation processes in heterogeneous systems. The high dependence of copolymer composition on
hydrodynamic conditions serves as evidence for the above statement.
In this case copolymer composition is affected by the reactivity of
the initial monomers and also by other physical and chemical properties of monomers, which determine the rate of their mass transfer
from the bulk into the reaction zones. Mass transfer processes in
cases of polymer synthesis in heterogeneous systems are numerous.
Some of them will be discussed below.

1. *The Mass Transfer of Monomers Through the Copolymer Film as a Limiting Stage*

Films of copolymer formed once can offer a substantial resistance to
the penetration of monomers from the bulk phase into the reaction zone.
Copolycondensation may be limited by the diffusion of monomers from
phase I into phase II through this polymeric film (Fig. 8.2). The
rate of diffusion of monomers through this polymeric film was shown
[24, 25] to be expressed by the following equation:

$$\frac{d[a]}{dt} = \frac{D^* K_D^A \, S'\{[A_o] - [a]\}}{\delta} \tag{8.14}$$

where D^* is the effective diffusion coefficient of the monomer through
the polymeric film, S' is the specific area (area divided by volume of
phase I), K_D^A is the distribution coefficient of monomer between phase
II and phase I (if reaction proceeds in phase I, $K_D = 1$), and δ is
the thickness of the polymeric film formed at a given time.

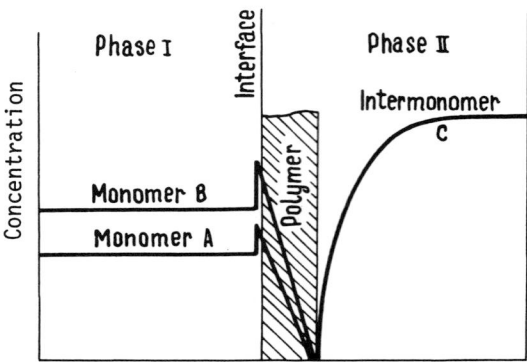

FIG. 8.2. Diagram of the copolycondensation process defined by the rate of mass transfer of monomers through the polymeric film.

If accumulation of monomers does not occur in reaction phase II and the monomers completely include into reaction in the narrow reaction zone of phase II near the interface, then Eq. (8.15) can be constructed using Eq. (8.14) for each of the monomers.

$$\frac{a}{b} \approx \frac{d[a]}{d[b]} = \frac{D_A^* K_D^A \{[A_o] - [a]\}}{D_A^* K_D^B \{[B_o] - [b]\}} \qquad (8.15)$$

This case is described and considered in Ref. [13].

An example is the copolycondensation of 1,4-tetramethylenediamine (A) and 1,10-decamethylenediamine (B) in the presence of isophthaloyl chloride in a chloroform-water system under static conditions (i.e., without stirring). In this case the diamines first diffuse from the water phase into the chloroform phase where the polymeric film is formed. They further penetrate through the polymeric film into the reaction zone which travels with the growth of the film. Copolymer composition for these conditions and experimental values of r are given in Table 8.4 [13].

To prove the diffusion nature of copolycondensation in the systems (see Table 8.4) it is necessary that the experimental value

TABLE 8.4

Copolymer Compositions and Copolycondensation Parameters of Tetramethylenediamine (A)-Decamethylenediamine (B)-Isophthaloyl Chloride in Chloroform-Water Systems under Static Conditions

Molar fraction A		Time of synthesis (min)	Monomer conversion B (%)	η_{sp} in H_2SO_4 0.5 g/dl	$r \pm \Delta r$
In initial mixture	In copolymer				
0.919	0.66	5	41	0.28	0.14 ± 0.04
0.919	0.73	5	34	0.43	0.19 ± 0.06
0.919	0.65	30	48	0.33	0.13 ± 0.04
0.919	0.60	30	47	0.57	0.09 ± 0.02
0.919	0.90	30	55	0.49	0.21 ± 0.09
0.919	0.74	60	51	0.44	0.21 ± 0.09
0.919	0.81	60	42	0.49	0.22 ± 0.10

of the copolycondensation parameter, r, and the r' value estimated from Eq. (8.15) coincide.

The diffusion coefficients of diamine through the polymeric film in the chloroform-water system at 25°C and the distribution coefficients of diamine (see Table 8.4) in this system were used for the estimation of r', the former for tetramethylenediamine and decamethylenediamine being $1.8-3.6 \times 10^{-8}$ cm^2/sec and 0.9×10^{-9} cm^2/sec, respectively [24]. The value of the copolycondensation parameter, r, calculated from Eq. (8.15) was 0.07-0.14. This value of r is close to the estimated r' thus confirming the diffusion mechanism of copolycondensation under static conditions.

The copolycondensation region determined by the monomer diffusion through the polymeric film has been observed with other monomer systems such as the ethylenediamine-phenylenediamine-adipoyl chloride system [11, 26].

A similar expression can be derived for copolycondensation occurring under dynamic conditions, i.e., with stirring. In this case

Eq. (8.14) includes the radius of a spherical drop [27], which does not alter Eq. (8.15).

It follows from the foregoing that the coefficients of monomer diffusion through the polymeric film may be determined from values of the copolycondensation parameter provided that the diffusion coefficient for one of the monomers is known.

The density of the homopolymeric film can also be evaluated because the diffusion coefficient is substantially determined by the film density. It is necessary for such evaluations that the monomers do not differ significantly in their reactivity.

2. The Mass Transfer of Monomers to the Reaction Zone as a Limiting Stage

The simplest case of copolymer formation is one in which the reaction rate is limited by the diffusion of monomers within one phase [28]. This takes place, for example, in the copolycondensation of two acid chlorides with one diamine when both acid chlorides diffuse into the reaction zone which resides in the organic phase and the intermonomer (diamine) diffuses into the reaction zone from the aqueous phase (see Fig. 8.3). The convection flows in the phases are indicated in

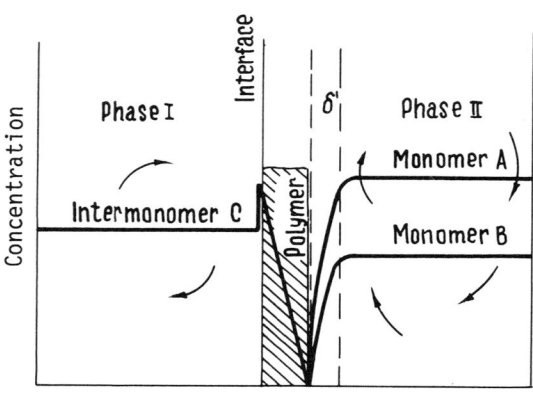

FIG. 8.3. Diagram of the copolycondensation process defined by the diffusion of monomers within one phase.

Fig. 8.3 by arrows; δ' stands for the thickness of the reduced liquid film which exhibits the main resistance to diffusion [29-31]. In this case the copolymer composition is determined by the competition between the diffusion currents of monomers A and B.

The equation for the copolymer composition is constructed from the equation of monomer diffusion, applied to each comonomer [29-31].

$$\frac{d[a]}{d[b]} = \frac{D_A([A_o] - [a])}{D_B([B_o] - [b])} \qquad (8.16)$$

where D_A, D_B are the diffusion coefficients of the monomers A and B in the same phase.

It is seen from Eq. (8.16) that the copolymer composition would be solely described by the diffusion coefficients and concentrations of monomers in the phase and would not depend on the density of the polymeric film or the reactivity of monomers.

The diffusion process appears to take place in the following systems: hexamethylenediamine-isophthaloyl chloride and adipoyl chloride [12]; hexamethylenediamine-adipoyl, sebacoyl, isophthaloyl, and terephthaloyl chlorides in pair with 4,6-dibromoisophthaloyl chloride [12], and possibly in other monomer systems [10].

In copolycondensation with aromatic acid chlorides, r = 1.05 [12] for such systems as hexamethylenediamine-isophthaloyl and 4,6-dibromoisophthaloyl chlorides; while for hexamethylenediamine-terephthaloyl and 4,6-dibromoisophthaloyl chlorides r = 1.17 [12]. The above copolycondensation parameters may correspond to the ratio of the monomer diffusion coefficients in the organic phase. The expressed coefficient ratios for those monomers [16] correspond to the observed copolycondensation parameters.

In processes involving readily hydrolyzable acid chlorides (as comonomers) the copolymerization parameter is affected by the hydrolysis, and therefore the obtained values of the above parameter should be subjected to refinement, for example, for systems given in Ref. 12.

The other type of copolycondensation process to be discussed in this section *is the process limited by mass transfer of monomers from*

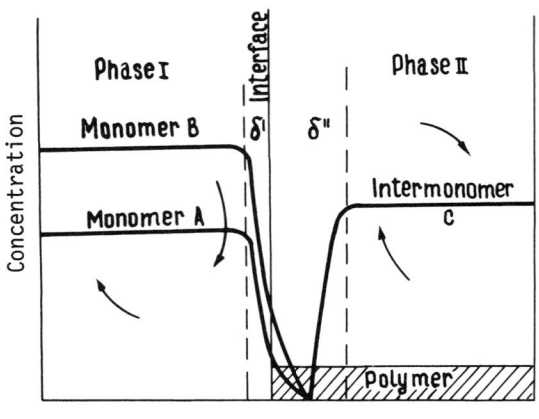

FIG. 8.4. Diagram of the copolycondensation process defined by the rate of mass transfer of monomers from one phase to another. Convection in phases is indicated by arrows and δ', δ'' are reduced liquid film which show main resistance to mass transfer [29-31].

one phase to another (Fig. 8.4). This is observed in interfacial copolycondensations when the polymeric film is absent and the rates of the propagation reactions are very high.

The following equation [32-34] is applicable in cases of mass transfer processes complicated by a chemical reaction:

$$\frac{d[a]}{dt} = K_M^A K_D^A S'\{[A_o] - [a]\} \tag{8.17}$$

where K_M^A is the coefficient of mass transfer involving the chemical reaction. The analogous equation may be written for the monomer B. Solving jointly the mass transfer equations for monomers A and B the following equation is obtained for the copolymer composition:

$$\frac{d[a]}{d[b]} = \frac{K_M^A K_D^A\{[A_o] - [a]\}}{K_M^B K_D^B\{[B_o] - [b]\}} \tag{8.18}$$

The above mechanism of copolycondensation is observed for the tetramethylenediamine-decamethylenediamine-isophthaloyl chloride system at the chloroform-water interface under the following conditions: intensive stirring, the total initial concentration of

8. Copolycondensation and Macroscopic Kinetics

diamines in the aqueous phase being 0.04 mol/liter, and the initial concentration of isophthaloyl chloride in chloroform 0.02-0.10 mol/liter. The comparison of the estimated value of $r' = 0.68$-0.83 with the experimental value of the copolycondensation parameter $r \lesssim 0.82$ confirms that the reaction is limited by the rate of diamine mass transfer.

The calculation of r' was carried out using values of $K_M^A \times K_D^A$ and $K_M^B \times K_D^B$ determined by model experiments with [13, 33] benzoic acid instead of isophthaloyl chloride.

The above process may be supposed for other systems of monomers having relatively high reactivity in processes involving stirring.

3. Adsorption of Monomers onto the Reaction Surface as a Limiting Stage

This case is most likely for the copolycondensation of highly surface-active monomers with high reactivity. The equation derived below from Eq. (8.13) describes the above copolycondensation situation:

$$\frac{d[a]}{d[b]} \approx \frac{b_A\{[A_o] - a\}^2}{b_B\{[B_o] - b\}^2} \approx \frac{\Gamma_A\{[A_o] - [a]\}}{\Gamma_B\{[B_o] - [b]\}} \qquad (8.19)$$

where b_A, b_B are the adsorption coefficients of monomers onto the interface and Γ_A, Γ_B are the specific adsorptions of monomers onto the interface.

Equation (8.19) was derived taking into consideration the adsorption-desorption equilibrium of monomers and reaction products as well as their diffusion and kinetic characteristics. The values Γ_A, Γ_B can be determined [35] from measurements of the surface tension of monomer solutions of different concentrations according to Gibb's equation [36]. The coefficients b_A, b_B are determined from the same experimental data utilizing Shishkovski's equation [37],

$$\sigma = \sigma_o - \alpha RT \ln\{1 + b_A[A]\} \qquad (8.20)$$

where σ_o is the interfacial tension, σ is the interfacial tension in presence of the monomer A, and α is the coefficient.

Another equation describing the copolymer composition for copolycondensation of surface active aliphatic homologues can be obtained from Eq. (8.19) and (8.20):

$$\frac{d[a]}{d[b]} = r_o L^{n_A - n_B} \frac{[A_o] - [a]}{[B_o] - [b]} \qquad (8.21)$$

where L is Traube's constant specifying the relative surface activity of homologues differing only in the number of methylene groups $(n_A - n_B)$ [36].

The case under consideration (in which adsorption is the rate-determining step) was found to occur in the copolycondensation of 1,4-tetramethylenediamine (A) and 1,10-decamethylenediamine (B) under static conditions and at low total concentrations of diamines (~ 0.02 mol.liter). The experimental values of the copolycondensation parameter are shown in Fig. 8.5 (by points), and the solid curve shows the estimated ratio of specific diamine adsorption as a function of the ratio of the two diamine monomers in the initial mixture. Figure 8.5 shows that the variation of the copolycondensation parameter and its numerical value agree well with the calculated ratio of specific diamine adsorptions, i.e., with the value $r' = \Gamma_A/\Gamma_B$ obtained from data given in Ref. 35. Therefore, a limitation by diamine adsorption onto the interface occurs under these conditions.

The limitation of copolymer formation by adsorption of diamines onto the interface at low concentrations also occurs for the synthesis of polydecamethylene isophthalamide, with the decamethylene concentration being less than 0.02 mol/liter. The polymer content in the

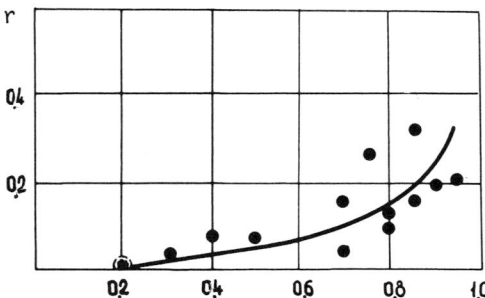

Molar fraction of A in initial mixture

FIG. 8.5. Relation of copolycondensation parameters-experimental points to ratio of adsorption activity of diamines curve Γ_A/Γ_B.

8. Copolycondensation and Macroscopic Kinetics

volume of "undisturbed" polydecamethylene isophthalamide film was investigated under static conditions as a function of decamethylenediamine concentration (< 0.016 mol/liter). The polymer content in the membrane at the interface determined the quantity of decamethylenediamine entering the reaction with isophthaloyl chloride in the reaction zone. It was found (see Fig. 8.6) to be linearly dependent upon the specific decamethylenediamine adsorption at the chloroform-water interface [13].

The limitation by diamine adsorption probably also takes place in the copolycondensation processes of ethylenediamine and hexamethylenediamine with adipoyl chloride in a benzene-water system without stirring when monomer and intermonomer concentrations are 0.05 mol/liter [11]. The experimental value of r = 0.27 found for this system is in close agreement with the specific adsorption ratio of diamines. The ratio of specific adsorption of ethylenediamine and hexamethylenediamine is 0.23 as estimated from the work of CH_2-group adsorption onto the chloroform-water interface [35] according to Traube's constant.

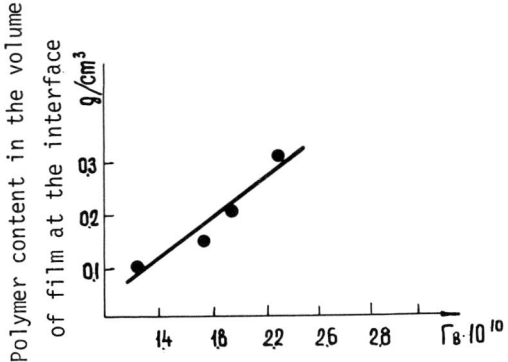

FIG. 8.6. Relation of polymer content in polydecamethylene isophthalamide film to specific adsorption of decamethylenediamine at the chloroform-water interface (Γ_B).

The copolycondensation in liquid-vapor interfacial systems also proceeds under the conditions considered above. The copolymer composition for these systems is described by Eq. (8.21) (see Chap. 7).

Thus, the limitation of copolycondensation by adsorption is common for heterogeneous two-phase systems. Equations (8.19) and (8.21) will also be true for the limitation of copolycondensation by desorption since the accumulation of the reaction products at the interface due to a slow rate of desorption leads to the limitation of the process as a result of restricting the monomer adsorption.

C. Copolymer Composition in Transitional Copolycondensation Regions

The above discussed macroscopic processes of interfacial copolycondensation may be observed even in a single system as experimental conditions vary. Changes in stirring intensity of monomer concentration, for instance, can affect unequally the rates of processes which determine the copolymer composition.

The intermonomer concentration does not appear in the equation describing the copolymer composition (Eq. 8.2), and its variation should not affect the copolycondensation parameter and copolymer composition. However, this influence is very sharp. This is well illustrated by the data of Fig. 8.7 which shows the variation of the experimental copolycondensation parameter, r, as the intermonomer concentration is varied in the copolycondensation of tetramethylenediamine (A)-decamethylenediamine (B)-isophthaloyl chloride.

The dependence of r on the intermonomer concentration in Fig. 8.7 conventionally can be considered as comprised of four regions a, b, c, d, with constant values. In accordance with discussion above the regions are as follows: a is the internal kinetic process, b is the diffusion process (a process limited by mass transfer), and c is also a diffusion process (where the process is limited by diffusion of monomers through the polymer film).

In addition, the curve in Fig. 8.7 contains regions p and q in which the value of r changes. These are called transitional regions.

8. Copolycondensation and Macroscopic Kinetics

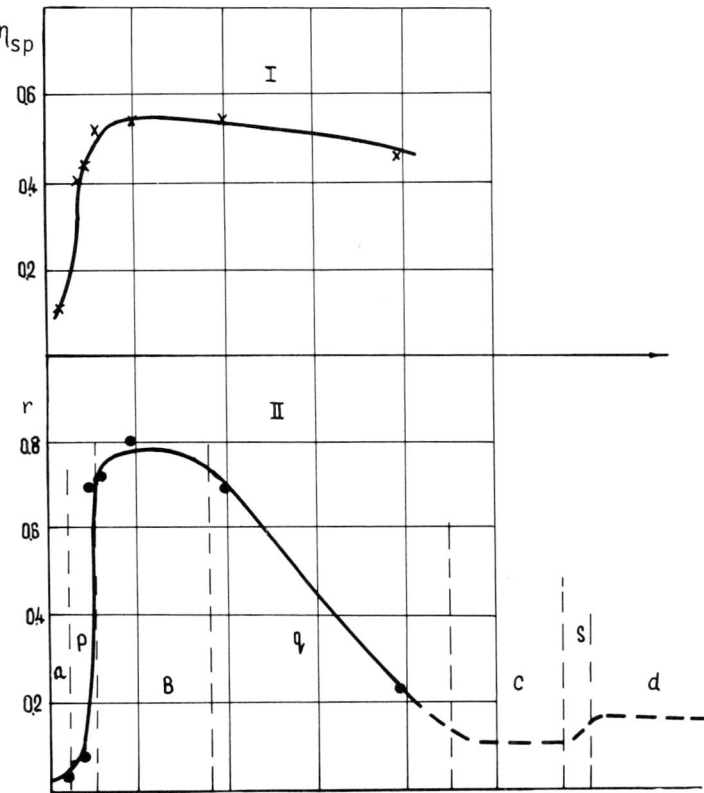

FIG. 8.7. Dependence of molecular weight (I) and copolycondensation parameter (II) on concentration of intermonomer (isophthaloyl chloride) in synthesizing copolyalkylene isophthalamides: a, b, c, d—main copolycondensation processes, p, q, s—transitional regions; η_{sp} (at conc. = 0.5 g/dl).

The transition from a kinetic process a into a diffusion process b, as observed in the region p, occurs very sharply.

This transition is related to the increasing rate of the chemical reaction as a result of the increase in isophthaloyl chloride concentration. The sharp transition from the kinetic process into a diffusion process was observed previously in processes with monofunctional compounds using intensive stirring [38, 39]. The transitional region q is due to decrease of the mass transfer coefficient as a result of increase in density and thickness of the polymeric film.

Regions d and s may be observed on further increasing the concentration of intermonomer; d is the process limited by the adsorption step, while s is the transitional region. These regions are observed very clearly in copolycondensation under static conditions.

The macroscopic process of copolycondensation may be changed not only by a variation of monomer concentrations but also by changing other conditions, such as stirring intensity, temperature, and catalysts. It is possible to change the region by varying solvent nature and the acid acceptor type [39a, 39b]. For example, the transition from process c into process d is expressed by the following ratio of variable parameters:

$$D^* \frac{K_D[A]}{\delta} = K^S \Gamma_A[c] \qquad (8.22)$$

In the case in which $D^* \{K_D[A]\}/\delta > K^S \Gamma_A [c]$ the reaction takes place by process c. This relation may be used for the determination of the rate constant of a heterogeneous reaction, K^S, by changing the inequality sign in region S. In accordance with calculations described in Ref. 13 a value of K^S of 2.4×10^4 liter/(mol sec) is calculated for the decamethylenediamine-isophthaloyl chloride system at the chloroform-water interface.

In addition to evidence about the macroscopic conditions, curves of the type shown in Fig. 8.7 allow one to obtain other information. Thus, from the value of the intermonomer concentrations in the transitional region p for various diacid chlorides, a series of their relative activities may be found.

D. *Copolymer Composition under Other Possible Copolycondensation Circumstances*

The above discussed conditions of interfacial copolycondensation in heterogeneous liquid systems does not cover the whole variety of macroscopic copolycondensation kinetics.

8. Copolycondensation and Macroscopic Kinetics

Thus, in interfacial copolycondensation the case may occur in which each of the monomers reacts by its own process. There may be cases in which the macroscopic process of a single reactant can vary during the course of reaction. Such is the case involving monomers with groups having considerably different reactivities in the copolycondensation such as of aminophenols, cis-2,5-dimethylpiperazine with acid halides. The interaction of the 4-amino group of cis-2,5-dimethylpiperazine follows a process determined by the mass transfer rate and the interaction of the 1-amino group of 2,5-dimethylpiperazine, having a lower reactivity because of steric hindrances from methyl groups, takes place by the kinetic process.

In addition, different reactions forming parts of a copolycondensation process may each occur under different variants of macroscopic kinetics. For example, in the case of the copolycondensation of two acid halides, copolymer formation may take place under conditions determined by diffusion rates of acid chlorides in the organic phase but their hydrolysis is limited by chemical reaction rate and proceeds also in the aqueous phase. Further, the polymeric film and its absorption of comonomers may strongly affect the macroscopic process.

The foregoing presents evidence from a wide variety of copolycondensations in heterogeneous liquid systems. Also, the synthetic processes of copolyesterification offers regions characteristic of only these processes. Therefore, when discussing any interfacial copolycondensation one should always determine and define the specific process of polycondensation.

IV. EFFECTS OF SIDE REACTIONS ON COPOLYMER COMPOSITION

Besides macromolecule formation in interfacial copolycondensation, side reactions of monomers usually take place.

As a rule, monomers participate in two types of such reactions—concurrent reactions with solvent or impurities, and consecutive-concurrent reactions of monomers with a byproduct, formed as a result of the polymerization reaction.

The water hydrolysis of acid chlorides, the formation of cyclic products of the monomers with an intermonomer, as well as some other reactions characteristic of known monomers, such as the oxidation of bisphenols by air, can be considered as side reactions of the first type proceeding simultaneously with the propagation reactions.

In all macroscopic regions concurrent reactions affect the copolymer composition and should be taken into account. An example of a homogeneous system serves to illustrate the effects of side reactions as occurs during the copolycondensation of two easily hydrolyzable acid chlorides.

The copolycondensation proceeds in the presence of a great excess of both an intermonomer (diamines) and a hydrolyzing agent. This case is described by the equation

$$\left[\frac{A_o - a_1(1 + r_A^*)}{A_o}\right] = \left[\frac{B_o - b_1(1 + r_B^*)}{B_o}\right]^{r_o[(1+r_A^*)/(1+r_B^*)]} \quad (8.23)$$

where A_o, B_o are initial amounts of A and B; a_1, b_1 are the amounts of monomers A and B incorporated into the polymer chain; r_o is the constant of relative activity; and r_A^*, r_B^* are the ratios of the monomer hydrolysis rates to the rates of incorporating the monomer into the copolymer for the monomers A and B, respectively. For pronounced hydrolysis r_A^* and r_B^* are large, while in the absence of hydrolysis r_A^*, $r_B^* = 0$.

Analysis of Eq. (8.23) shows that in the case of intensive side reactions (e.g., hydrolysis) polymer composition is determined not only by the relative reaction rates of monomers A and B with the intermonomer but also by the relative intensity of their hydrolysis. This conclusion is qualitatively confirmed by experimental results [15, 47a].

In the instance of weak hydrolysis (where $r_A^* \approx 0$, $r_B^* \approx 0$) Eq. (8.23) becomes identical to Eq. (8.8). The values of r_A^* and r_B^* may be evaluated from the yields of the homopolymers [20].

The hydrochloride-forming reaction in the interfacial synthesis of polyamides from diamines and acid chlorides, and in the synthesis

8. Copolycondensation and Macroscopic Kinetics

of polyurethanes from diamines and bis-chloroformates can be related to consecutive-concurrent side reactions.

Since the hydrochloride-forming reaction proceeds very rapidly ($\sim$R-NH$_2$ + HCl $\longrightarrow$ $\sim$R-NH$_2 \cdot$ HCl) the concentration of HCl in the presence of diamine is assumed to be near zero.

The differential equation describing copolymer composition for copolycondensation processes in the absence of an acceptor for HCl is as follows:

$$\frac{d[a]}{d[b]} = r \frac{\{[A_o] - 2[a]\}}{\{[B_o] - 2[b]\}} \tag{8.24}$$

The factor 2 shows that during incorporation of monomer A into the polymer, "double" monomer consumption takes place as a result of both polymer and salt formation.

Equation (8.24) is applicable for the calculation of the copolycondensation parameter based on the data of copolymer analysis for monomers of similar reactivity, or for systems in which the ratio of the constants of reaction rate of monomers with the intermonomer is equal to the ratio of the rate constants for the reaction with a hydrochloride. Polymer composition can be expressed by the equation (8.2, 8.7) if, in the copolycondensation of two diamines with a diacid chloride as the intermonomer, an HCl acceptor stronger than diamines is used such that all evolved HCl instantly combines with it.

All the above examples relate to effects of the side reaction of the monomers on copolymer composition during copolycondensation in the kinetic process.

In the case of copolycondensation in diffusion processes, a similar approach, i.e., calculation of decrease in monomer concentrations in the two processes, is also applicable, but the expression of final solutions for these systems is very complicated.

If the intermonomer participates in side reactions, the copolymer composition generally does not depend on the course of such side reactions, with negligible if any degree of monomer conversion.

Other reactions can occur in addition to those mentioned above. These include reactions of chain-breaking, branching (cross-linking), and reactions resulting in the formation of polymers (i.e., polymers containing several types of connecting linkages). The chain-breaking and branching reactions do not change the copolymer composition, but they may produce a change of monomer distribution throughout the chain.

Relations resulting in the formation of homopolymers containing several types of connecting linkages can take place during the synthesis of copolymers. For instance, the reaction of diisocyanates with water during the synthesis of polyurethanes results in the formation of urea units in the chain. Essentially such synthesis of homopolymers may be considered as a real copolycondensation. That is why in the interbipolycondensation process such reactions cause multicomponent copolycondensations.

V. NONUNIFORMITY OF COPOLYMER COMPOSITION UNDER DIFFERENT VARIANTS OF MACROSCOPIC KINETICS

In considering composition nonuniformity of the condensation copolymers one should first consider its variation with the degree of conversion.

Copolymers of uniform composition should be obtained in copolycondensations within the kinetic region, provided that the comonomers used are of equal reactivity [18]. When copolycondensation of monomers having different reactivities takes place within the kinetic region, the composition nonuniformity of the copolymers changes with conversion until in the extreme case homopolymer is obtained. In this case the content of the more reactive monomer in the initially formed copolymer is higher than in the copolymer formed at the end of the reaction.

Turska [40] recently reported that nonuniform copolyesters were obtained in spite of the fact that the reactivities of initial monomers of 2,2-bis(4-aminophenyl)propane and of 4,4'-dihydroxybinaphthyl-1,1', i.e.,

8. Copolycondensation and Macroscopic Kinetics

in the reaction with terephthaloyl chloride differed only by a factor of two. Reaction was effected using the solution polycondensation method. Diffusion, adsorption, and other characteristics of monomers together with some degree of conversion are of great importance in nonkinetic processes in which comonomers of equal reactivities undergo reaction. Therefore, nonuniform copolymers may have been obtained even with comonomers of equal reactivities. Hence, the composition nonuniformity of copolymer can be changed with changing the macroscopic processes and the reaction site.

The interface copolycondensation of resorcinol and m-xylylene diamine with isophthaloyl chloride (intermonomer) depended on solvent nature and also on acid acceptor [39a, 39b]. Acid acceptor greatly influenced the structure of resulting copolymers. The formation of polyamides was accelerated in the presence of a strong base, while a weak base increased the ester formation.

Investigation of the composition nonuniformity of copolyisophthalamides utilizing tetramethylene- and decamethylenediamines in different variants of macroscopic kinetics and with low degrees of conversion has been reported [13, 14]. The analysis of thermomechanic and turbidimetric curves showed that random copolymers were formed by the process defined by the rate of diamine mass transfer, and that copolymers of nonuniform composition were formed by the process where reaction was limited by the rates of diffusion of tetramethylenediamine and decamethylenediamine through the polymeric film.

Thus, analysis of composition nonuniformity of copolymers in interfacial copolycondensation has to be carried out taking into account macroscopic kinetics, with degree of monomer conversion being low.

VI. PRACTICAL METHODS AND CONDITIONS OF VARIOUS COPOLYMER SYNTHESES

It is clear from the foregoing that many characteristics of copolymers obtained from interfacial copolycondensation are determined by the macroscopic process.

The molecular weight of copolymers depends on the process. This is well illustrated with the data of Fig. 8.7. The specific viscosity of copolyamides is seen to repeat the change in the copolycondensation parameter.

The background of copolymer synthesis in interfacial polycondensation is analogous to that of homopolymer syntheses reported in Refs. 20 and 41. Features related to the copolymer synthesis are discussed below.

Some initial data are necessary for the purpose of obtaining copolymers with specific characteristics (composition, etc.). Such minimum data include the value of the relative activity of the monomers, r_o, and the value of the copolycondensation parameter, r.

The relative activity of monomers can be evaluated from the physical and chemical characteristics related to the reactivity dissociation constants [42], spectral characteristics [43, 44], various thermochemical evidence [45], from correlation equations of different types [46, 47], or calculated by a quantum-mechanical method [39c]. When the above data is not available it is necessary to carry out model experiments, i.e., to estimate the relative yield of products resulting from the reaction of comonomers with an intermonomer or its monofunctional analogs.

It is advisable in practice to carry out preliminary experiments in order to estimate the value of r. This can be done by determining copolymer composition with a fixed composition of the initial mixture. Since in copolycondensation, unlike vinyl copolymerization, the dependence of copolymer composition on the initial mixture composition is single valued, the value r is given approximately by a single point on the curve of copolymer composition vs. initial mixture composition.

Selecting the monomer pairs. The same copolymer can be obtained from different monomers; for example, copolymers containing different monomer acid fragments in the chain can be produced from various monomers such as acid chlorides, fluorides, and others. It is advisable to use monomers of equal activities in the kinetic process to obtain statistically uniform copolymers. Statistically uniform

8. Copolycondensation and Macroscopic Kinetics

copolymers can be obtained in copolycondensation of monomers of different activity in the nonkinetic regions of macroscopic kinetics (this has been pointed out in the previous section, for example, in the region defined by the rate of monomer mass transfer). Producing random copolymer in the kinetic region is possible in the case of a single-stage process, where the unit distribution is close to the statistical one and depends on differences in comonomer reactivity [1c, 1b].

The initial composition of the comonomer mixture determines the copolymer composition in accordance with Eqs. (8.3) and (8.4) or by means of an alternative more complicated equation.

The relationship of quantities between comonomers and intermonomers does not affect the copolymer composition in the kinetic process but does substantially influence the copolymer molecular weight (see Sect. 8.III).

Sharp differences in the reactivities of monomers or functional groups of monomers leads to alternation [48, 49] or possibly to a block structure of polymer, or a mixture of homopolymers [41, 50, 51].

The concentration relationship in the kinetic process affects both the copolymer composition and molecular weight, which serves as a useful tool for the researcher.

The degree of conversion determined by the relationship between the quantities of comonomers and intermonomers is, as a rule, different for each reagent. It determines practically all characteristics of the copolymer (see Sect. 8.V).

Nature of organic phase. The copolymer composition and other characteristics can be considerably affected by the nature of the organic phase [39a, 47a, 52]. Therefore, in addition to the above initial data it is very important to consider also such characteristics as distribution factors, chemical and adsorption activity of monomers, etc., which vary with the nature of the organic phase.

Stirring does not affect the copolymer composition and other characteristics in the kinetic processes, but it may influence the copolymer characteristics through changing the macroscopic process.

Emulsifiers. No experimental data is available on the influence of emulsifiers on copolycondensation.

Versions of procedure. The order of reagent addition is an important method of controlling the copolymer chain structure. It has been reported previously [41] that copolymers of block structure may be obtained by addition of the reagents in a given sequence.

A modification of the sequential addition utilizes a two-stage synthesis. For example, the synthesis of trimer having a specific composition may be carried out first, followed by condensing with another monomer. This method is used, for example, in obtaining polyamidesters of regular structure [48, 53, 53a] and polyurethane carbonates [54-56].

Sometimes the method of gradual feeding of one comonomer is used as a variation of the method described above [57, 57a]. Taking into account the particular macroscopic process, the method of gradual feeding can be used for obtaining uniform copolymers from monomers with widely different activities [58]. By proceeding continuously, the above process also offers advantages in productivity.

The experimental data presently available still do not allow one to obtain a complete and detailed picture of the macroscopic kinetics of interfacial polycondensation. As knowledge of macroscopic kinetics and the physical and chemical characteristics of interfacial systems accumulates, investigators will become more confident of the syntheses of copolymers with predetermined properties.

REFERENCES

1. *Copolymerization, High Polymers Series* (J. E. Ham, ed.), Vol. 1, Interscience, New York, 1964.
1a. V. A. Vasnev and S. I. Kuchanov, *Usp. Khim.* (Russ.), *42*(12): 2194 (1973).
1b. S. V. Vinogradova, V. A. Vasnev, and Yu. I. Perfilov, *Acta Chim. Acad. Sci. Hung.*, *81*(2):209 (1974).
1c. V. V. Korshak, S. V. Vinogradova, S. I. Kuchanov, V. A. Vasnev, G. D. Markova, and A. I. Tarasov, *Vysokomol. Soedin.*, *Ser. A*, *16*(9):1992 (1974).

8. Copolycondensation and Macroscopic Kinetics

1d. V. V. Korshak, S. V. Vinogradova, V. A. Vasnev, Yu. I. Perfilov, and P. D. Okulevich, *J. Polym. Sci., Polym. Chem. Ed.* 11(9):2209 (1973).

2. V. Z. Nikonov, L. B. Sokolov, G. B. Babur, Yu. V. Sharikov, and E. A. Emelin, *Vysokomol. Soedin., Ser. A,* 11(4):673 (1969).

3. D. A. Frank-Kamenetskii, *Diffusion and Heat Tranference in Chemical Kinetics* (Russ.), Moscow, Ed. Nauka, 1967.

4. O. V. Smirnova, G. S. Kolesnikov, A. K. Mikitaev, and Salman Takhir Salman, *Vysokomol. Soedin., Ser. A,* 9(9):1989 (1967).

5. O. V. Smirnova, G. S. Kolesnikov, A. K. Mikitaev, and T. G. Krichevskaya, *Vysokomol. Soedin., Ser. A,* 10(1):96 (1968).

6. G. S. Kolesnikov, O. V. Smirnova, and A. K. Mikitaev, *Vysokomol. Soedin., Ser. B,* 12:449 (1970).

7. O. V. Smirnova, A. K. Mikitaev, *Vysokomol. Soedin., Ser. B,* 12(6):445 (1970).

8. G. S. Kolesnikov, O. V. Smirnova, and A. K. Mikitaev, *Vysokomol. Soedin., Ser. A,* 9(3):684 (1967).

9. G. S. Kolesnikov, O. V. Smirnova, and Sh. A. Samsoniya, *Vysokomol. Soedin., Ser. B,* 9:49 (1967).

10. L. B. Sokolov and T. L. Kruglova, *Vysokomol. Soedin.,* 2:704 (1960).

11. V. V. Korshak, T. M. Frunze, and L. V. Koslov, *Isv. Akad. Nauk SSSR, Otd. Khim. Nauk* (Russ.), 2227 (1962).

12. V. V. Korshak, T. M. Frunze, and L. V. Koslov, *Isv. Akad. Nauk SSSR, Otd. Khim. Nauk* (Russ.), 2227 (1962).

13. V. Z. Nikonov, Dissertation, Moskow, 1968.

14. L. B. Sokolov, V. Z. Nikonov, and G. N. Shilyakova, *Vysokomol. Soedin., Ser. A,* 11(3):616 (1969).

15. V. V. Korshak, T. M. Frunze, and L. V. Koslov, *Vysokomol. Soedin.,* 2:838 (1960).

16. S. Bretshneider, *Properties of Liquid and Gases* (Russ.), Moscow, Ed. Inlit, 1968.

17. G. A. Streichman, *Zh. Prikl. Khim.,* 32:673 (1959).

18. L. Beste, *J. Polym. Sci.,* 36:313 (1959).

19. V. Tynzye, *J. Polym Sci., Part A-1,* 3(10):3654 (1965).

20. L. B. Sokolov, *Synthesis of Polymers by Polycondensation,* Israel Program for Scientific Translation, Jerusalem, 1968.

21. L. B. Sokolov and S. S. Medved, *Vysokomol. Soedin., Ser. B,* 10(7):514 (1968).

22. M. I. Temkin, *Zh. Fiz. Khim.,* 11:169 (1938).

23. S. L. Kiperman, *Introduction in Kinetics of Heterogeneous Catalytic Reactions* (Russ.), Ed. Nauka, Moscow, 1964, p. 134.

24. V. Z. Nikonov, L. B. Sokolov, *Vysokomol. Soedin., Ser. B.*, 10(5):337 (1968).
25. V. Z. Nikonov, G. N. Kudryavtseva, *Vysokomol. Soedin., Ser. A,* 14(6):1216 (1972).
26. L. V. Koslov, Dissertation, Moscow, 1963.
27. J. H. Stanley, *A.I.Ch.E. J.*, 13(7):6, 1124 (1967).
28. J. Astarita, *Mass Transfer with Chemical Reaction,* Elsevier, Amsterdam, Ch. 10 (1967).
29. V. G. Levich, *Usp. Khim.* (Russ.), 34(11):1846 (1965).
30. Yu. A. Byevich, *Zh. Fiz. Khim.* (Russ.), 39(4):829 (1965).
31. R. Treybal, *Liquid Extraction,* 2 ed., McGraw-Hill, New York.
32. P. V. Danckverts, *Ind. Eng. Chem.*, 43(6):1460 (1951).
33. A. A. Abramson and M. V. Ostrovskii, *Zh. Prikl. Khim.* (Russ.), 36(3):608 (1963).
34. M. Kh. Kishinevskii and T. S. Kornienko, *Zh. Prikl. Khim.* (Russ.), 37:1285 (1964).
35. V. Z. Nikonov, L. B. Sokolov, *Zh. Fiz. Khim.* (Russ.), 43(4):1032 (1969).
36. N. K. Adam, *The Physics and Chemistry of Surface,* Oxford University, London, p. 375 (1938).
37. *Kratkayaa khimicheskaya enzyclopediya* (Russ.), Moscow, 4:97 (1965).
38. A. A. Abramson, *Zh. Prikl. Khim.* (Russ.), 37(8):1771 (1964).
39. V. G. Levich, A. M. Brodskii, *Dokl. Akad. Nauk SSSR,* 165(3):607 (1965).
39a. N. Ogata and T. Ikari, *J. Polym. Sci., Polym. Chem. Ed.,* 11(8):1939 (1973).
39b. N. Ogata, *Kobunshi, High Polymer Japan,* 22(1):36 (1973).
39c. V. A. Kosobuckii, V. K. Belyakov, and S. G. Tumakov, in *Quantum Chemistry* (Russ.), K-125, Kishinev, Shtinca (1975).
40. E. Turska, *Vysokomol. Soedin., Ser. A,* 15(2):393 (1973).
41. P. W. Morgan, *Condensations Polymers by Low Temperature and Interfacial Methods,* Part V, Interscience, New York, 1967.
42. A. Albert and E. Serzhent, *Ionization Constants of Acids and Bases* (Russ.), Inlit, Moscow-Leningrad, 1964.
43. R. Zbinden, *Infrared Spectroscopy of High Polymers,* Academic, New York, 1964.
44. P. L. Luisi, E. Chielline, F. F. Franchini, and M. Orienti, *Macromol. Chem.,* 112:197 (1968).

8. Copolycondensation and Macroscopic Kinetics

45. D. F. Sokolova and L. B. Sokolov, *Vysokomol. Soedin., Ser. A,* 14(4):894 (1972).
46. V. A. Palm, *Foundations of Quantitative Theory of Organic Reactions* (Russ.), Ed. Khimiya, Leningrad, 1967, p. 72.
47. P. Sykes, *Guidebook to Mechanism of Organic Chemistry,* Longman, London, 1967.
47a. T. Kivotsukurei, T. Banba, and Y. Nishikawa, *Kobushi Kagaku,* 30(342):587 (1973).
48. Y. Preston, *J. Polym. Sci., Part A-1,* 8(11):3135 (1970).
49. W. Y. Bailey and I. Okamoto, *Polymers Preprints,* 12(1):177 (1971).
50. O. V. Smirnova and S. A. Samsonova, *Vysokomol. Soedin., Part B,* 12(5):384 (1970).
51. Brit. Patent 1,210,596.
52. K. Tsuyoshi and K. Akie, *Kobunshi Kagaku,* 28(312):279 (1971).
53. T. Kiyotsukuri and Y. Shimomura, *Kobunshi Kagaku,* 28(314):516 (1971).
53a. G. Lenz, H. Krim, and H. Schnell, U.S. Patent 3,809,679 (1974).
54. C. Jiori, *Polymer Preprints,* 11:326 (1970).
55. C. Jiori, *Polymer Preprints,* 11:281 (1970).
56. G. S. Kolesnikov, Patent SSSR, 221, 281, 1967.
57. L. B. Sokolov, *Vysokomol. Soedin., Ser. A,* 13(6):1425 (1971).
57a. S. I. Kuchanov, *Vysokomol. Soedin., Ser. B,* 16(2):136 (1974).
58. K. Ikeda and Y. Sekine, *Ind. Eng. Chem. Prod. Res. Develop.,* 12(3):202 (1973).

9.

INTERFACIAL, COLLOIDAL, AND KINETIC ASPECTS OF EMULSION POLYMERIZATION

John L. Gardon

Research Department
Coatings and Ink Division
M&T Chemicals, Inc.
Southfield, Michigan

I.	INTRODUCTION	206
II.	GENERAL FEATURES OF EMULSION POLYMERIZATION	206
	A. Typical Recipes and Results	206
	B. Description of a Simple Mechanistic Model	207
III.	THEORETICAL PREDICTION OF PARTICLE SIZE, CONVERSION RATE, AND MOLECULAR WEIGHT	208
	A. The Rate of Particle Nucleation	208
	B. The General Equation for Conversion Rate	209
	C. Description of Interval I	210
	D. Description of Interval II—When Termination is Instantaneous	211
	E. The Consequences of Noninstantaneous Termination During Interval II	212
	F. Description of Interval III	215
IV.	EXPERIMENTAL TESTS FOR THE THEORY	216
	A. Parameters Characterizing Typical Reaction Systems	216
	B. Particle Size and Particle Number	219
	C. The Conversion-Time Curve	225
	D. Molecular Weight	232
	E. Particle Size Distribution	234

V.	ALTERNATE MODELS FOR PARTICLE FORMATION	235
	A. The Harkins Hypotheses	235
	B. Particle Formation in the Absence of Micellar Surfactants	236
	C. Particle Nucleation at Submicellar Surfactant Concentrations	240
	D. Critique of the Present Theories of Particle Nucleation	241
VI.	SWELLING OF LATEX PARTICLES BY MONOMERS	242
	A. The Thermodynamics of Swelling	243
	B. Monomer Concentration in Latex Particles During Emulsion Polymerization	248
	REFERENCES	249

I. INTRODUCTION

In emulsion polymerization, kinetic, colloidal, and interfacial phenomena interact in a complex manner. For certain model systems, the mechanism is well understood and theoretical predictions are in good quantitative accord with experimental data. On the other hand, there is a large body of experimental results that is as yet not properly understood.

In this chapter the colloidal and interfacial aspects of emulsion polymerization are discussed, with emphasis on monomers that fit into the framework of the Smith-Ewart and succeeding theories [1-8]. Good general reviews on emulsion polymerization are available elsewhere [9-15].

II. GENERAL FEATURES OF EMULSION POLYMERIZATION

A. Typical Recipes and Results

A monomer such as methyl methacrylate or styrene is blended with an aqueous surfactant solution containing, for example, 0.5 to 1% sodium lauryl sulfate. The monomer/water ratio is 30/70 to 60/40. The temperature is kept constant at 40 to 80°C and the reaction is

9. *Aspects of Emulsion Polymerization*

initiated by addition of 0.05 to 1% of potassium persulfate.*
Generally, the reaction proceeds rapidly, at a few percent per
minute conversion rate, and yields a polymer of several million
molecular weight in the form of a colloidal dispersion with a
particle size in the 0.1 µm range.

B. *Description of a Simple Mechanistic Model*

Initially the monomer is present at two loci, in the monomer-swollen
micelles of about .005 µm diameter and in the monomer droplets of
1 to 10 µm diameter [10]. The total surface area of the micelles is
much larger than that of the droplets so that the radicals generated
in the aqueous phase enter droplets to a neglible extent; almost all
radicals enter micelles and initiate polymerization in them. A
micelle "stung" by a radical becomes a latex particle. The growing
latex particles absorb monomer from other micelles and from monomer
droplets, and the monomer/polymer ratio within the particles is kept
at a level corresponding to the thermodynamic equilibrium. The
growing particles also adsorb surfactants from as yet "unstung"
micelles to keep the particle surfaces saturated with surfactant.
The nucleation of particles stops when all micellar surfactant is
consumed. From this point the particle number remains constant.
It is convenient to designate the particle nucleating phase of
emulsion polymerization as Interval I.

The locus of polymerization is solely in the monomer-swollen
latex particle. No significant polymerization can take place in
water, since the polymer produced is insoluble in water. Polymerization in the monomer droplets is insignificant since their total

---

*Such experiment can be performed in a three-neck flask under
nitrogen blanket with mild stirring. The stirring rate has no
effect on the results as long as it is adequate for keeping the
aqueous phase saturated with monomer and is not too vigorous to
cause coagulation by shear.

surface area is much smaller than that of the combined micelles and
latex particles; and the rate of radical absorption is proportional
to the surface area of the absorbing unit.* Consequently, droplets
will not absorb a significant number of polymer-initiating radicals.

Even if the monomer is a good solvent for the polymer, the latex
particles can absorb only a limited amount of monomer for reasons
explained in Sect. 9.VI. The monomer diffuses to the polymerizing
particles much faster than it is consumed by polymerization. Consequently,
the polymerizing particles are kept saturated with monomer
as long as there is enough unconverted monomer present as monomer
droplets. In Interval II the particles are saturated with monomer
as they are in Interval I. The difference between Intervals I and
II is that in Interval I the particle number is increasing, while in
Interval II it is constant.

As the reaction proceeds further, the amount of unconverted
monomer becomes insufficient to keep the growing particles saturated
with monomer. At this stage, the reaction enters Interval III. Here
the particle number is constant and the monomer concentration in the
particles decreases with increasing conversion.

III. THEORETICAL PREDICTION OF PARTICLE SIZE, CONVERSION RATE, AND MOLECULAR WEIGHT

A. The Rate of Particle Nucleation

The surfactant and initiator influence the outcome of the results
through parameters S and R. The surfactant molecules contained in
1 cm^3 water occupy a total surface area of S cm^2 at the water/organic
phase interface, and the number of radicals produced per second in
1 cm^3 of water is R. Both parameters are time-independent constants.
This constancy rests on the implicit assumptions that the area per
soap molecule is the same at the particle/water and micelle/water
interfaces, and that the half-life of the initiator is much longer

---

*This point is clarified in Sect. 9.V.B.

9. Aspects of Emulsion Polymerization

than the duration of the experiment. These two parameters are calculable by Eqs. (9.1) and (9.2) from the Avogadro number, N_A, the molar surfactant and initiator concentrations [S] and [I], the area per surfactant molecule, A_s, and the decomposition rate constant of the initiator, k_d.

$$S = N_A A_s [S] \qquad (9.1)$$

$$R = 2 N_A k_d [I] \qquad (9.2)$$

Initially, all R radicals enter into micelles, and the rate of particle nucleation, dN_t/dt, equals R. As new particles are formed, they compete with the micelles for radical capture. It is assumed that these particles are efficiently covered by surfactant molecules, so that during Interval I the interfacial area between the water and the organic phase is constant and equals S. The surface of the latex particles is $4\Pi \Sigma n_i r_i^2$, where n_i is the number of particles with radius r_i. The fraction of radicals which do not nucleate new particles is $(4\Pi/S) \Sigma n_i r_i^2$. Consequently, the rate of particle nucleation is defined by Eq. (9.3).

$$\frac{dN_t}{dt} = R(1 - \frac{4\Pi}{S}\Sigma n_i r_i^2) \qquad (9.3)$$

For solving this equation the growth rate of particles must be defined so that the time dependence of $\Sigma n_i r_i^2$ can be established.

B. The General Equation for Conversion Rate

In homogeneous polymerization, the volume of the reaction mixture is approximately constant and the time dependence of the molar concentrations of monomer and radical, [M] and [R], can be described in terms of the propagation rate constant, k_p:

$$\frac{-d[M]}{dt} = k_p [M][R] \qquad (9.4)$$

During Intervals I and II of emulsion polymerization the monomer concentration at the locus of the reaction, in the particles, is approximately constant (see Sect. VI) but the volume of the reaction site increases with time. It is convenient to express conversion in terms of P, the volume of polymer present in unit volume of water.

The volume of the reaction site, that is of all the monomer swollen particles, is given by Eq. (9.5) in terms of ϕ_m, the constant monomer volume fraction in the particles during Intervals I and II.

$$\frac{P}{1 - \phi_m} = \frac{4\Pi}{3}\Sigma n_i r_i^3 \approx \frac{4\Pi}{3} N_t r^3 \tag{9.5}$$

The molar volume of the monomer is V_m, Q is the average number of radicals per particle, and d_p and d_m are the densities of polymer and monomer. The following relationships hold by definition:

$$-d[M] = \frac{d_p}{d_m V_m} \left(\frac{1 - \phi_m}{P}\right) dP \tag{9.6}$$

$$[M] = \frac{\phi_m}{V_m} \tag{9.7}$$

$$[R] = \frac{N_t Q}{N_A} \left(\frac{1 - \phi_m}{P}\right) \tag{9.8}$$

Combination of Eqs. (9.4) to (9.8) gives the conversion rate for emulsion polymerization:

$$\frac{dP}{dt} = \frac{k_p}{N_A} \left(\frac{d_m}{d_p}\right) N_t \phi_m Q \tag{9.9}$$

C. Description of Interval I

The value of Q in Eq. (9.9) has been the subject of extensive investigations [7, 16, 17]. In Interval I each particle initially contains one radical; later radicals enter into existing particles and the termination and reinitiation processes cause Q to change. It was found [7] that the mathematical approximation $Q = 1$ does not lead to significant errors for describing Interval I even if it is mechanistically incorrect. Accordingly, the rate of volume growth in Interval I can be approximated by combining Eqs. (9.5) and (9.9) and setting $Q = 1$. The resulting equation, Eq. (9.10) defines a new parameter,

9. *Aspects of Emulsion Polymerization*

K, which can be used in solving the Eq. (9.3) of particle nucleation by mathematical methods described by Gardon [5, 7] and Smith [1].

$$K = \frac{3}{4\pi} \frac{k_p d_m}{N_A d_p} \frac{1-\phi_m}{\phi_m} = \frac{dr^3}{dt} \qquad (9.10)$$

The resulting experimentally verifiable equations, Eqs. (9.11), (9.12), and (9.13), give N, the final number of particles per cubic centimeter of water; r_{rmc}, the root-mean-cube average particle radius* after polymerization is complete at a monomer/water ratio of m/w; and P_{cr}, the volume of polymer formed per unit volume of water at the time when particle nucleation is completed (i.e., when the surface of all monomer-swollen particles becomes equal to S).

$$N = 0.208 S^{0.6} \left(\frac{R}{K}\right)^{0.4} \qquad (9.11)$$

$$r_{rmc} = 1.05 \left(\frac{m}{w}\right)^{0.333} \left(\frac{d_{water}}{d_p}\right)^{0.333} S^{-0.2} \left(\frac{K}{R}\right)^{0.133} \qquad (9.12)$$

$$P_{cr} = 0.209 S^{1.2} \left(\frac{K}{R}\right)^{0.2} (1 - \phi_m) \qquad (9.13)$$

D. *Description of Interval II—When Termination Is Instantaneous*

For calculating rates and molecular weights, the value of Q of Eq. (9.9) must be defined. During Interval II all other parameters in Eq. (9.9) are constant, so that if Q is a constant the conversion rate must also be constant.

The simplest assumption involves Smith-Ewart's Case 2. If the particles are small so that the particle number is large and if the initiation rate is low, then N/R, the average interval between the entry of two radicals into a single particle, will be relatively large. If a radical enters a particle already containing one radical,

---

*$r_{rmc} = (\Sigma n_i r_i^3 / \Sigma n_i)^{1/3}$. It should be noted that r_{rmc} and N are interdependent. The following equation holds: $(4\pi/3) r_{rmc}^3 = (d_{water}/d_p)(m/w)/N$.

the time needed for cross-termination of the two radicals is much shorter than N/R so that the particle will be radical-free until a third radical enters. Consequently, each particle contains one radical half of the time and no radical the other half of the time, and Q becomes equal to 0.5. The Smith-Ewart rate, B, for Interval II is derived from Eqs. (9.9), (9.10) and (9.11) with Q = 0.5 and N_t = N. The final result is shown in the equivalent Eqs. (9.14) and (9.15).

$$\frac{dP}{dt} = B = 0.5 \frac{k_p}{N_A} \frac{d_m}{d_p} \phi_m N \tag{9.14}$$

$$B = 0.185 \frac{(k_p \phi_m S d_m)^{0.6}}{(d_p N_A)^{0.6}[R(1 - \phi_m)]^{0.4}} \tag{9.15}$$

If termination is instantaneous, each chain grows for an average time equal to the interval between the entry of two radicals into a particle, N/R. During this interval, the polymer volume is produced by each growing chain at a rate of 2B/N. It follows that in Interval II the molecular weight should be the constant Smith-Ewart value, M_{se} given in Eqs. (9.16), (9.17) and (9.18).

$$M_{se} = \frac{2B}{N} \frac{N}{R} N_A d_p = \frac{2BN_A d_p}{R} \tag{9.16}$$

$$= \frac{k_p \phi_m N d_m}{R} \tag{9.17}$$

$$= 0.318[N_A d_p(1 - \phi_m)]^{0.4} (\frac{d_m k_p \phi_m S}{R})^{0.6} \tag{9.18}$$

E. *Consequences of Noninstantaneous Termination During Interval II*

If N/R is comparable to the time needed for termination, several radicals can coexist in each particle so that Q of Eq. (9.9) and dP/dt will vary with time. A convenient form of Eq. (9.9) for this case is Eq. (9.19):

$$\frac{dP}{dt} = 2BQ \tag{9.19}$$

9. Aspects of Emulsion Polymerization

For calculating Q, a useful approximation valid for methyl methacrylate and styrene polymerization is that all radicals become terminated within particles, there is no termination between particles, and radicals do not desorb from particles. The termination rate within the particles is influenced by the time-independent termination rate constant, k_t.* For homogeneous polymerization, Eq. (9.20) describes the variation of the radical concentration [R] with time:

$$\frac{d[R]}{dt} = 2k_d[I] - k_t[R]^2 \tag{9.20}$$

This equation can be transformed [3, 6] to one useful for emulsion polymerization in a manner analogous to the derivation of Eq. (9.9) from Eq. (9.4). The new formula, Eq. (9.21), contains—instead of the radical concentration, [R]—the number of radicals in an individual particle, q:

$$\frac{dq}{dt} = \frac{R}{N} - \frac{k_t(1 - \phi_m)N}{PN_A} q(q - 1) \tag{9.21}$$

Equations (9.22) and (9.23) below are valid by definition. They contain f_q, the fraction of particles containing q radicals.

$$\Sigma f_q = 1 \tag{9.22}$$

$$Q = \Sigma q f_q = f_1 + 2f_2 + 3f_3 + 4f_4 + \cdots \tag{9.23}$$

The system is defined by Eqs. (9.9), (9.21), (9.22), and (9.23). An interesting consequence of the rather involved mathematical derivations [6] is that the distribution of radicals among particles is a unique function of Q, shown in Fig. 9.1. Since every particle

---

*The value of k_t is known to be dependent mainly on the viscosity of the reaction medium. This viscosity is approximately constant during Interval II because the monomer/polymer ratio in the particles and the molecular weight of the polymer vary little.

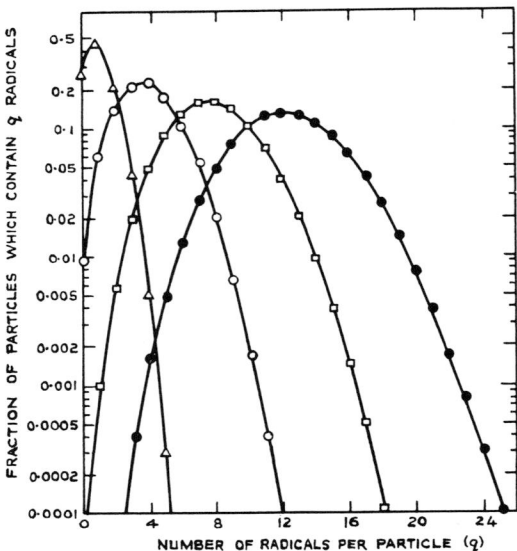

FIG. 9.1. Theoretical distribution of radicals among particles at different average number of radicals per particle Q. Since the growth rate of each particle is proportional to q, each curve also gives the distribution of growth rates per particle [6].

grows proportionally to q, situations leading to Q > 0.5 necessarily cause broadening of the particle size distribution according to the theoretical distribution curves of Fig. 9.1.

The effects of slow termination rates can be conveniently expressed [6] by a new parameter, A, defined in Eq. (9.24).

$$A = 0.102 \frac{k_p^{1.94}}{k_t^{0.94}} \frac{d_m}{d_p N_A} \frac{\phi_m^{1.94}}{(1-\phi_m)^{0.94} R} \qquad (9.24)$$

The mathematics for solving Eqs. (9.19) and (9.21) to (9.24) are given by Stockmayer [3], Gardon [6], Ugelstad and Mark [8], and Alexander and Napper [9]. While there are subtle differences in the various solutions, these are not significant. Equations (9.25) to (9.28) give the results for conversion, average number of radicals

9. Aspects of Emulsion Polymerization

per particle, and number or weight average molecular weights in Gardon's [6] formalism.

$$P = At^2 + Bt \tag{9.25}$$

$$Q = 0.5(1 + \frac{4A}{B^2}P)^{0.5} \tag{9.26}$$

$$M_n = \frac{(4AN_A d_p/BR)P}{[1 + (4A/B^2)P]^{0.5} - 1} \tag{9.27}$$

$$M_v = \frac{(B^3 N_A d_p/3AR)\{[1 + (4A/B^2)P]^{1.5} - 1\}}{P} \tag{9.28}$$

Evidently, the assumption of instantaneous termination used in Sect. 9.III.D becomes valid only at low initiator and high surfactant concentrations, when Eq. (9.29) is applicable:

$$\frac{4A}{B^2}P \ll 1 \tag{9.29}$$

The theory described here accounts only for positive deviations from the Smith-Ewart Case 2, i.e., when $Q > 0.5$. This theory satisfactorily describes, as will be shown, the emulsion polymerization of methyl methacrylate, styrene, and butyl acrylate.

In the polymerization of vinyl chloride, Q is found [8] to assume values much lower than 0.5. The reason is that radicals do not cross-terminate solely within isolated latex particles. They can desorb from the particles, and growing chains located in the different particles can cross-terminate. A theoretical description is given by Stockmayer [3], O'Toole [4], and Ugelstad and Mark [8], but it is not described here in detail.

F. Description of Interval III

After the monomer droplets disappear, all unconverted monomer is in the particles if the solubility of monomer in water is low. In this Interval III, the monomer volume fraction, ϕ_{eff}, in the particles is not constant but decreases with increasing time. The viscosity in the particles increases so that k_t decreases. The shape of the conversion-time curve cannot be predicted because the dependence of k_t on ϕ_{eff} is not known.

The experimental value of ϕ_{eff} is calculable from the fractional conversion, C, if the solubility of the monomer in water is low. The shape of the conversion-time curve defines Q and also the molecular weight. The following equations can be used to describe Interval III:

$$\phi_{eff} = \frac{1-C}{1-C[1-(d_m/d_p)]} \qquad (9.30)$$

$$Q = \frac{dP}{dt} \frac{\phi_m}{2\phi_{eff}B} \qquad (9.31)$$

$$M_n = \frac{2d_p N_A}{R} \frac{P}{t} \qquad (9.32)$$

IV. EXPERIMENTAL TESTS FOR THE THEORY

A. Parameters Characterizing Typical Reaction Systems

The experimentally verifiable equations of the theory contain no adjustable constants. All the parameters can be determined from measurements not involving emulsion polymerization. The precision of the theory can be tested by comparing the results of emulsion polymerization experiments with the theoretical predictions calculated with the aid of such independently established parameters. Before describing experimental results, the values of these parameters are discussed in Sect. 9.IV.A.

1. The Initiator Parameter, R

The number of radicals produced in 1 cm^3 water per second, R, is readily calculable from the known decomposition rate constant [18] of potassium persulfate according to Eq. (9.2). The values of R,

9. Aspects of Emulsion Polymerization

for a 1% aqueous solution and pH 3.5, are 1.4×10^{13}, 5.8×10^{13}, and 2.45×10^{14} at 40°, 50°, and 60°C, respectively.*

2. The Soap Parameter, S

The soap molecules present in 1 cm^2 occupy S cm^2 area at the interface. This soap parameter is calculable from Eq. (9.1) if the area per soap molecule, A_s, is known. The values of A_s can be determined by one of three methods; by X-ray measurements, by applying the Gibbs absorption isotherm to the concentration dependence of interfacial tension, or by soap titration of latexes with known particle sizes. In the latter method, the latex is prepared with the aid of the surfactant whose A_s value is to be determined. After the polymerization is complete, more surfactant is added. As long as the surface of the latex particles is incompletely covered by the surfactant, the surface tension remains constant. On subsequent surfactant addition, the surface tension decreases. The various methods are generally in agreement within 20 to 30%. The value of S is also somewhat temperature dependent [19] and may vary with the adsorbant surface [20]. Useful values of S are shown in Table 9.1 for 1% aqueous surfactant solutions. Evidently, the applicable values of R and S for any known concentration are readily calculated from the data shown here for 1% concentration.

It is important to note that the efficiency of the various surfactants for reducing the particle size is best measured by the parameter S, as defined in Table 9.1. Generally, commercially useful latexes are obtained when the value of S is about 10^5 cm^2/cm^3. Table 9.1 shows that sodium lauryl sulfate is about 6 times more efficient

---

*The decomposition rate of $K_2S_2O_8$ is quite sensitive to pH. For unbuffered systems a pH of about 3.5 is a good representative value because $K_2S_2O_8$ yields acidic $KHSO_4$ as a byproduct of its decomposition.

TABLE 9.1

The Value of the Soap Parameter S for
1% Aqueous Solutions Calculated from Eq. (9.1)

Solution	$S \times 10^{-5}$	Ref. for A_s
Na lauryl sulfate	1.25	21, 22
K laurate	1.26	23
Na laurate	1.06	24
Na myristate	0.77	24, 25
Na palmitate	0.53	24
Na oleate	0.50	24
Na stearate	0.46	24
Na decylbenzene sulfonate	0.59	26
Adduct of octylphenol and 40 ethylene oxide units	0.18	21

than nonionic surfactants with long ethylene oxide chains. Typical commercial recipes contain either 0.5 to 1% of the former or 3 to 6% of the latter.

3. The Volume Growth Rate Constant, K

For calculating the value of K by Eq. (9.10), the propagation constant, k_p, determined by homogeneous photopolymerization, is used [28, 29]. The values in units of cm^3/(mole sec) are 8.3×10^4, 1.24×10^5, and 1.76×10^5 for styrene at 40°, 50°, and 60°C, respectively. The corresponding values for methyl methacrylate are 1.95×10^5, 2.70×10^5, and 3.6×10^5. The ratio of polymer-to-monomer densities are 1.12 and 1.27, respectively, for these two monomers. The volume fraction of the monomer in the latex particles at saturation swelling, ϕ_m, generally depends on the surfactant used, as discussed in Sect. 9.VI. However, only a small error is committed by using typical ϕ_m values of 0.6 and 0.73 for styrene and methyl methacrylate, respectively.

9. *Asepcts of Emulsion Polymerization*

The primary parameters given above define K values at 40°, 50°, and 60°C of 4.2×10^{-20}, 6.3×10^{-20}, and 8.84×10^{-20} for styrene and 1.64×10^{-19}, 2.28×10^{-19}, and 3.08×10^{-14} for methyl methacrylate. The units of K are cm^3/sec.

4. The Conversion Rate Parameters, A and B

The conversion rate at instantaneous termination during Interval II can be measured as the volume of polymer per unit volume of water per second. This value, B, is calculated from Eq. (9.15) with the aid of the parameters defined above: R, S, K, k_p, ϕ_m, and d_p/d_m.

Often it is convenient to express the slope of the linear component of the conversion-time curve in percent/minute units. The corresponding value, b, is calculable by Eq. (9.33) from B, from the monomer/water ratio, and from the polymer density.

$$B = 1.667 \times 10^{-4} \left(\frac{m}{w}\right) \left(\frac{d_{water}}{d_p}\right) b \qquad (9.33)$$

For calculating A, the value of the termination rate constant, k_t, should be known from independent experiments. In the very viscous medium inside the latex particles, the value of k_t is several orders of magnitude lower than generally determined at infinite dilution. Theoretically, the lowest limits of the ratio k_t/k_p in an infinitely viscous medium is about 12 [30, 31, 32]. As will be shown, the values of A determined from emulsion polymerization generally correspond to a k_t/k_p ratio of the order of 100, and this ratio is in good agreement with values independently determined for very viscous reaction media.

B. Particle Size and Particle Number

1. Values at Full Conversion

Tables 9.2, 9.3, and 9.4 show that the theoretical predictions for particle sizes or particle numbers can be very accurate for styrene and methyl methacrylate monomers, potassium persulfate initiators, and three different anionic surfactants. Of particular interest is Table 9.2, which shows that the accuracy of experimental predictions

TABLE 9.2

Final Particle Radii Obtained with Methyl Methacrylate[a]

$r \times 10^6$ (cm)			Concentration in water (%)		
Theory (Eq. 9.12)	Electron microscopy	Light scattering	Soap	Initiator	Salt
2.4	3.2	—	4.88	1.54	0
2.8	3.6	—	3.44	0.905	0
3.6	4.3	—	2.00	0.035	0
5.2	5.1	—	0.514	0.206	0
6.2	7.35	—	0.323	0.107	0
6.2	5.2	—	0.302	0.117	0
6.2	5.25	6.8	0.242	0.164	0
7.85	6.3	7.9	0.256	0.029	0
7.85	7.1	8.7	0.256	0.029	0.05
7.85	—	8.6	0.256	0.029	0.15
7.85	8.25	9.9	0.256	0.029	0.30
7.85	8.8	—	0.192	0.044	0
7.85	8.15	—	0.158	0.059	0
7.85	8.5	—	0.128	0.081	0
9.9	8.85	—	0.203	0.0065	0.05
9.9	8.65	9.5	0.152	0.010	0
9.9	—	9.6	0.152	0.010	0.025
9.9	8.75	10.3	0.152	0.010	0.05
9.9	9.55	—	0.100	0.0184	0
12.42	10.05	—	0.12	0.0023	0.05
12.42	12.0	—	0.08	0.0042	0.05

[a] Procedure: soap, sodium lauryl sulfate; initiator, $K_2S_2O_8$; m/w = 40/60. Salt (Na_2SO_4) was added to low initiator recipes to reduce viscosity. All experiments were run in three-necked flasks, with gentle stirring, under nitrogen at 40°C [33].

TABLE 9.3
Final Particle Radii and Molecular Weights Obtained with Styrene[a]

Temperature (°C)	m/w	% of Water Soap	% of Water Initiator	$r \times 10^6$ (cm) Electron microscopy	$r \times 10^6$ (cm) Theory	MW $\times 10^{-6}$ Expt.	MW $\times 10^{-6}$ Theory (Eq. 9.18)
			Experiments by Gardon [33]				
40	40/60	0.63	1.39	3.0[b]	4.3	4.02	5.20
40	40/60	0.66	0.23	4.8	5.45	3.2	15.6
60	10/90	0.66	0.25	2.2	3.38	2.0	3.8
60	40/60	0.66	0.23	4.2	4.22	2.35	3.8
60	40/60	0.66	0.23	4.3	4.22	3.54	3.8
60	40/60	0.67	0.24	4.64	4.22	3.78	3.8
80	20/80	0.125	0.125	3.7	3.32	—	—
80	20/80	0.5	0.125	2.50	2.50	—	—
			Experiments by Smith [34]				
30.5	37/63	0.5	0.165	7.90	8.31	23	20
50	37/63	0.5	to	5.80	5.84	4.4	6.2
70	37/63	0.5	0.172	4.82	4.12	1.1	1.4
90	37/63	0.5		4.20	3.70	0.32	0.27
70	37/63	2.0	0.516 to 0.516	3.6	2.89	2.1	4.5
70	37/63	0.5		4.53	3.78	0.78	0.57

[a] The initiator was always $K_2S_2O_8$. The soap in Gardon's experiments was sodium lauryl sulfate; in those of Smith it was SF Flakes. The latter soap is known to be a mixture of sodium stearate, oleate, and palmitate, and an S value of 5×10^4 is applicable at 1% concentration. Smith did not calculate theoretical particle sizes and molecular weights since at the time of his work the presently available theoretical equations were not known.

[b] This value of r was determined by soap titration, not by electron microscopy.

TABLE 9.4

Particle Number and Linear
Interval II Rate Obtained with Styrene[a]

Derivation	N per cm^3 of water	Conversion rate (%/min)
Experiment	7.6×10^{14}	0.97
Theory (Eq. 9.11)	4.7×10^{14}	—
Theory (Eqs. 9.15 and 9.33)	—	0.79

[a]Data by Burnett and Lehrle [35] at m/w = 12/88, run with $K_2S_2O_8$ and potassium stearate at 0.075% and 1.25% concentrations, respectively, and at 40°C. The theoretical data are newly calculated.

is maintained when the surfactant concentration is varied 60-fold and initiator concentration is varied 140-fold, corresponding to a 5- or 125-fold variation of particle radius or particle number, respectively. Further experimental data showing good agreement with theoretical predictions are presented by Gardon [33] and Smith [36].

Unfortunately, not all available data agree with the theoretical predictions as well as those of Table 9.2, 9.3, and 9.4. The theory predicts particle numbers about 2 to 3 times too large or particle radii about 1.25 to 1.45 too small when compared to the styrene data of Bartholome et al. [37], Van der Hoff [38], Robb [39], and Hamada et al. [40]. Similar discrepancy applies to the particle sizes obtained with methyl methacrylate by Gerrens [41]. Van der Hoff [38] and Hamada et al. [40] attribute these findings to poor radical utilization efficiency. However, if the particle numbers were proportional to $(eR)^{0.4}$ instead of $R^{0.4}$ (Eq. 9.11), where e is the efficiency, the experimental findings would require e to be in the range of 0.05 to 0.18. This would also imply that the molecular weights calculated by assuming e = 1 should be 5 to 20 times too low. The data of Sect. 9.IV.D show that the experimental and theoretical molecular weights are similar if e = 1 is used in the

9. Aspects of Emulsion Polymerization

calculations. Consequently, it is unlikely that the discrepancies in particle sizes are caused by poor radical utilization efficiency.

Perhaps the theoretical model is inadequate* and can predict only the order of magnitude but not the exact value of the particle number. However, the very good correspondence between theory and experiment for a large body of data cannot be fortuitous. The possibility exists that those experimental data which differ from the theoretical predictions may have been subject to systematic errors. In the determination of average particle sizes, it often happens that the too-small particles become "lost."

2. The Constancy of Particle Numbers in Intervals II and III

The final particle size and particle number are discussed above. It is also of interest that the theory predicts constancy of particle numbers when the reaction proceeds beyond a conversion corresponding to P_{cr} of Eq. (9.13). The value of P_{cr} is insensitive to monomer and initiator properties being proportional to a very low, 1/5-th power of K/R. In most practical recipes, P_{cr} should have a very low value, of the order of 0.01. It follows that particle nucleation should stop at a few percent conversion if m/w > 30/70. Indeed, as predicted, constancy of particle number after a few percent conversion has been observed for the polymerization of styrene [14, 39], methyl methacrylate [31], methyl acrylate [42], and vinyl acetate [43].

In Table 9.5, some data are analyzed which were not designed for testing the theoretical prediction that the particle number should be independent of conversion in Intervals II and III. Indeed, these data predate the publication of the Smith-Ewart theory.

If the particle number is constant, the particle volume—that is, the cubed diameter—must be proportional to conversion, so that the ratio between the diameter and the cube root of conversion is constant.

---

*A critique of the nucleation model is presented in Sect. 9.V.

TABLE 9.5

Variation of Particle Size With
Conversion of Sytrene/Isoprene = 25/75[a]

Conversion = 100C (%)	3.0	5.3	8.5	12.2	14.5	17.6	22.4	27.7	40.3	55.5	62
Diameter = $2r$ (Å)	160	205	250	270	290	310	340	360	405	450	466
$2r/C^{1/3}$ (Å)	514	545	568	544	552	553	559	552	548	558	559
% Surfactant in water phase	60	37	21	12	<8	<8	<8	<8	<8	<8	<8

[a]Procedure—monomer/water: 36/54, 2.5% potassium myristate in water, 50°C, trace of persulfate and mercaptan. The particle size was determined by light scattering after removal of unconverted monomer. The first two rows are taken from Table I of Klevens [25], the third row is newly calculated. The fourth row is calculated from Fig. 1 of Klevens [25]. After 14.5% conversion, the surfactant level in water corresponds to the undesorbed soap at incomplete surface coverage.

9. *Aspects of Emulsion Polymerization*

This is demonstrated for the copolymerization of styrene with isoprene in Table 9.5.

C. *The Conversion-Time Curve*

The conversion rate is theoretically predicted [5, 7] and experimentally found to rise rapidly during Interval I [13, 37, 44, 45], level off to a constant value or increase slightly with time during Interval II, and decrease at the onset or during Interval III. Because of the low value of P_{cr}, the fine features of the curve during Interval I are missing in most available data. The data that are convenient to interpret from the theoretical standpoint relate to Interval II.

1. Linear Conversion-Time Relationship in Interval II

Numerous authors determined the linear Interval II rate, the particle number, and the monomer concentration in the particles and calculated the value of k_p from these experimental data according to Eq. (9.14). Satisfactory agreement with independently determined k_p values were thus obtained for styrene [14, 35, 36, 44, 47, 51] and methyl methacrylate [41]. For a number of other monomers shown in Table 9.6, the k_p values calculated by this method are of a plausible order of magnitude though direct comparison with independently determined values cannot be made because these are not available. It is likely that the Smith-Ewart Case 2 model (Sect. 9.III.D) is not grossly in error for these monomers.

It is possible to test simultaneously the Smith-Ewart nucleation model and their Case 2 for Interval II by calculating the linear Interval II rate from the primary parameters, k_p, $\emptyset_m$, S, and R, by Eqs. (9.15) and (9.33), without using the experimental particle number. Tables 9.4 and 9.7 show excellent agreement between theory and experiment.

TABLE 9.6

Propagation Constants Calculated from Particle Number, Monomer Concentration in Particles, and Linear Interval II Rate of Emulsion Polymerization

Monomer	Temp. for k_p (°C)	$k_p \times 10^{-5}$ [cm^3/(mol sec)]	Activation Energy for k_p (kcal/mol)	Ref.
Butadiene	60	1.0	9.3	52
Dimethylbutadiene	60	1.2	9.0	53
Isoprene	60	5.0	9.8	51
Vinyltoluene	60	3.36	13.4	51
o-Methylstyrene	60	1.38	13.9	51
p-Methylstyrene	60	2.18	7.66	51
Chloroprene	40	2.20	—	47

TABLE 9.7

Linear Interval II Conversion Rates for Styrene Polymerization[a]

Mol/liter		Conversion rate (%/min)	
Initiator	Soap	Expt.	Theory (Eqs. 9.15, 9.33)
3.24×10^{-3}	4×10^{-2}	1.09	1.06
1.62×10^{-3}	4×10^{-2}	0.83	0.80
0.81×10^{-3}	4×10^{-2}	0.63	0.61
0.40×10^{-3}	4×10^{-2}	0.45	0.46
0.20×10^{-3}	4×10^{-2}	0.33	0.35
1.62×10^{-3}	2×10^{-2}	0.54	0.53
1.62×10^{-3}	8×10^{-2}	1.29	1.21
1.62×10^{-3}	16×10^{-2}	1.93	1.84

[a]Data by Brietenbach et al. [52]. Procedure: 50°C, m/w = 25/75; initiator: α,α'-azobismethylbutyronitrile-γ-Na-sulfonate; soap: potassium palmitate. The rate constant of initiator decomposition was separately determined and was found to be 3.35×10^{-6} sec^{-1}.

9. Aspects of Emulsion Polymerization

2. Deviation From Smith-Ewart Case 2 in Interval II

It should be recalled that the linear relationship between conversion and time (P = Bt) can be obtained only at relatively low initiator and high surfactant concentrations. Under such conditions, the Bt term dominates the general conversion-time relationship ($P = At^2 + Bt$). The parameter B is proportional to the particle number; that is, to the 0.4th power of the initiator concentration, as shown in Eqs. (9.14) and (9.15). According to Eqs. (9.2) and (9.24), the parameter A is quite sensitive to the initiator concentration, being proportional to it, and is independent of soap concentration and of the particle number. It follows that if everything else is kept equal but the particle number is varied by salt addition, the conversion-time curve will be linear for small particle size latexes and convex to the time axis at a large particle size. This theoretical prediction is proved correct as shown by the data of Fig. 9.2. Similar sets of curves are expected and obtained if the soap concentration is varied while the initiator level is kept constant. The highest surfactant level gives a linear conversion-time curve with high slope. Reduction in surfactant level imparts curvature to the conversion-time curves and reduces the slope at a given conversion. Such families of curves were obtained for the monomers quoted in Fig. 9.3. Figure 9.3 shows that, for the largest particle sizes, the second-order term dominates the conversion-time curves ($P = At^2$). In this limiting case, the square root of conversion becomes linear with time.

Figure 9.4 presents conversion-time curves that can be conveniently analyzed from the theoretical point of view. The arrows indicate the end of Interval II. Table 9.8 gives the kinetic parameters and particle size data of several experiments similar to those of Fig. 9.4.

The parameters A and B were determined by least-square fit of the experimental data to Eq. (9.25). From these, and the experimental values of N and ϕ_m, the ratio of k_t/k_p was calculated using Eq. (9.34), which follows from Eqs. (9.14) and (9.24):

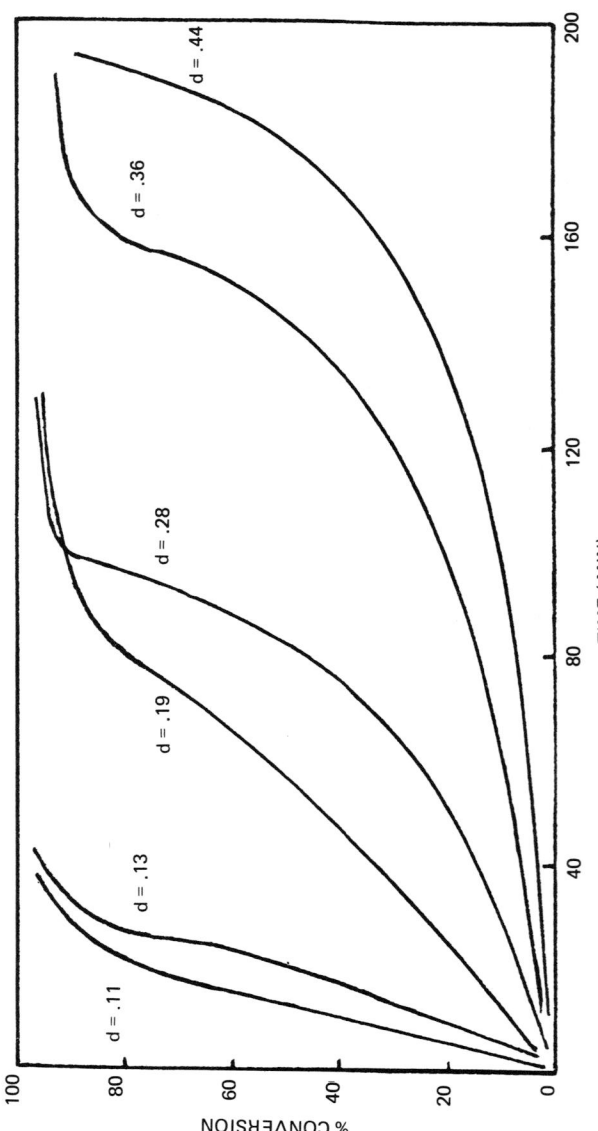

FIG. 9.2. Conversion-time curves obtained with methyl methacrylate at 60°C. The final particle diameters in microns are indicated and were determined by light scattering. The monomer/water ratio is 25/75. The initiator is $K_2S_2O_8$ at 0.05%, and the surfactant is Tergitol 7 at 0.09%. The particle size was varied by adding various amounts of $CaCl_2$ to the charge [53].

9. Aspects of Emulsion Polymerization

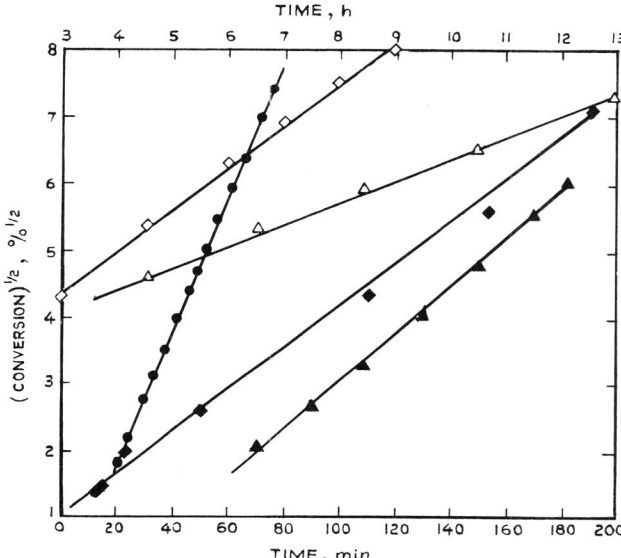

FIG. 9.3. Time dependence of the square root of conversion for relatively large particle-size emulsions: (△) butadiene, data of Wall et al. [54]; (◊) butadiene-styrene copolymer, data of Wall et al. [54]; (●) vinyl acetate, data of Napper and Parts [55]; (♦) vinylidene chloride, data of Hay et al. [56]; (▲) methyl methacrylate, data of Zimmt [53]. The time is expressed in hours for St/Bu and in minutes for VAc, VCl_2, and MMA. The data refer to the largest-size particle emulsion of each quoted publication. Taken from Gardon [57].

$$\frac{k_t}{k_p} = 0.158 \left(\frac{RB}{AN}\right)^{1.062} \frac{\phi_m}{1 - \phi_m} \tag{9.34}$$

It is noteworthy that the average number of radicals per particle, Q, is significantly larger than 0.5 throughout the reaction. The experimental B values of Table 9.8 are in satisfactory agreement with the theoretical values, similarly to the B values that were calculated from the linear conversion-time relationships, shown in Tables 9.6 and 9.7 where Q was 0.5.

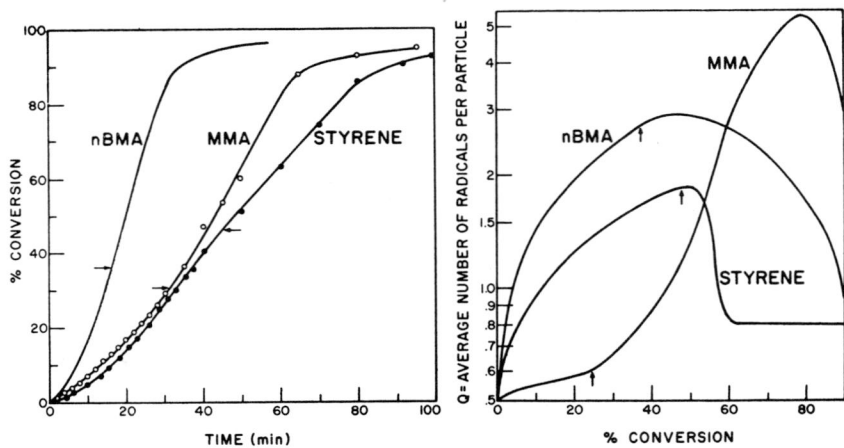

FIG. 9.4. Interdependence of conversion, time, and average number of radicals per particle. The arrows indicate where the monomer droplets disappear. The data with styrene and n-butyl acrylate were obtained at 60°C., those with methyl methacrylate at 55°C. Other experimental details are shown in Table 9.8. The values of Q are calculated from Eqs. (9.26) and (9.31) [57].

For proving the validity of the theoretical model, the parameter A, in addition to B, must be significant from the mechanistic standpoint. According to Table 9.8, the k_t/k_p ratios calculated with the aid of the experimental A values are of the order of 10 to 100. This range of k_t/k_p ratios is plausible for the very viscous medium existing within the latex particles. Similar ratios were obtained when Hummel et al. [45, 58] determined k_t and k_p directly, without involving any assumptions on the mechanism of emulsion polymerization. Hummel's experiments involved the study of the decay and increase of conversion rate when the polymerizing latex system was subjected to intermittant radiation. Also, North [59, 60] obtained similar k_t/k_p values in homogeneous polymerization in the presence of inert thickeners.

There are other data in the literature which show that, for styrene and methyl methacrylate, Q can be larger than 0.5 and that k_t/k_p is in the 10 to 100 range. These data are not independent

TABLE 9.8

Kinetic Data Obtained With Three Monomers[a]

Parameter	Methyl methacrylate		Styrene		Butyl acrylate
Procedure					
Temperature	40°C	55°C	50°C	60°C	60°C
Monomer/water	35/65	40/60	37/63	40/60	47/63
Na lauryl sulfate (%)	0.63	0.244	—	0.67	1.01
SF Flakes (%)	—	—	0.5	—	—
$K_2S_2O_8$ (%)	0.23	0.165	0.17	0.24	0.204
Results					
$r \times 10^6$ at full conversion (cm)					
Experiment [electron microscope]	7.8	5.25	5.8	4.6	5.21
Theory [Eq. 9.12]	5.75	6.2	5.8	4.2	—
$B \times 10^5$ (sec^{-1})					
Experiment	5.48	5.26	2.97	10.2	8.6
Theory [Eq. 9.15]	4.8	8.3	3.55	5.8	—
$A \times 10^9$ (sec^{-2}), experiment	2.29	19.4	1.52	10.9	187
k_t/k_p [Eq. 9.34]	77	41	150	141	9

Parameter	40°C MMA
Temperature	40°C
Monomer/water	38/62
Na lauryl sulfate (%)	1.47
$K_2S_2O_8$ (%)	0.142
Experiment r	4.9
Theory r	5.2
Experiment B	5.87
Theory B	6.5
A experiment	0.98
k_t/k_p	73

[a]Data of the 50°C styrene experiment from Smith [34], all other data from Gardon [57]. The experimental parameters A and B were determined by least-square fit to the data. Gardon's paper [57] presents more data of this nature. Other k_t/k_p values for methyl methacrylate are 29, 92, 100 and 115; for styrene 15; and for butyl acrylate 4 and 88.

confirmation of the theory of Sect. 9.IV because they involve the very similar theory of Stockmayer [3]. Nevertheless, it is worth recording that Gerrens [41, 46] and Van der Hoff [38, 61] presented data showing trends consistent with the present analysis.

D. *Molecular Weight*

Table 9.4 shows good agreement between the experimental molecular weight at full conversion and that calculated from theory for Interval II assuming instantaneous termination (M_{SE}). It is of interest that this simple form of the theory predicts the right order of magnitude of the molecular weights. Nevertheless, the good agreement in Table 9.4 must be fortuitous. Actually, the molecular weight produced in Interval II is higher than M_{SE} due to slow termination, and the molecular weight of the polymer of Interval III is lower than M_{SE} because of slow propagation in the particles containing decreasing amount of monomer. The material at full conversion contains both kinds of polymers.

Jovanovic et al [62] reported good agreement between M_{SE} and molecular weight obtained up to 37% conversion of styrene. Table 9.9 compares the conversion dependence of molecular weight in Gardon's [57] and Smith's [34] work to the theoretical predictions that take slow termination rate into account. The agreement is excellent for styrene, which terminates solely by recombination, as is implicitly assumed in deriving the theory. The lower than theoretical values experimentally obtained for methyl methacrylate are due to predominant termination by disproportionation which, if it were the sole termination mode, would lead to molecular weights one-half of those obtainable by recombination. The present theoretical and experimental data would agree well if 67% of the termination steps involved disproportionation and 33% involved combination. Independent data obtained in solution polymerization [63] suggest 82% and 18%, respectively. The discrepancy is within experimental error.

TABLE 9.9

Variation of Molecular Weight with Conversion in Styrene and Methyl Methacrylate Polymerization[a]

Monomer	Conversion, %	$\bar{M}_v \times 10^{-6}$		$\bar{M}_n \times 10^{-6}$, calculated	
		Exptl.	Calculated, Eq. (9.28)	Eq. (9.27)	Eq. (9.32)
Styrene (50°C)	5.1	2.66	3.85	3.49	—
	14.0	3.84	4.35	4.24	—
	31.9	4.68	4.90	4.72	—
	44.0	5.05	5.40	5.17	5.17
	61.0	5.38	—	—	5.62
	73.0	5.28	—	—	5.61
	80.5	5.12	—	—	4.62
Styrene (60°C)	3.7	1.53	2.34	2.29	—
	12.4	2.3	2.48	2.32	—
	21.2	2.9	2.59	2.54	—
	30.7	3.1	2.72	2.62	—
	38.7	3.3	2.82	2.70	—
Methyl methacrylate (55°C)	9.8	3.22	4.77	4.10	—
	12.75	3.38	5.22	4.80	—
	16.75	3.70	5.35	5.20	—
	21.40	3.70	5.96	5.35	5.35
	36.05	4.36	—	—	6.3
	59.85	4.65	—	—	7.0
	87.90	4.82	—	—	8.2
	93.3	4.65	—	—	7.25
	93.6	4.11	—	—	6.20

[a] The experimental details are given in the third, fourth and fifth columns of Table 9.8. For calculating the theoretical molecular weight by Eqs. (9.28) and (9.29) the experimental values of the rate parameters A and B were used.

E. *Particle Size Distribution*

Sundberg and Eliassen [64] calculated theoretical particle-size distributions for the case where termination is instantaneous and predicted about symmetrical distributions, similar to Gaussian, for particle volumes (or r^3). These distributions are predicted to become narrower with a smaller median value if the initiator concentration is increased. These predictions are consistent with the data published by Gerrens [13].

The theoretical size distribution for the case where slow termination affects the results has not yet been established. The theoretical distribution of growth rates in Fig. 9.1 suggests that, as the average number of radicals per particle increases with conversion, the particle size distribution must broaden and become skewed in such a way that the population becomes increasingly rich in large particles, and increasingly poor in small particles. The experimental results of Fig. 9.5 confirm this theoretical expectation.

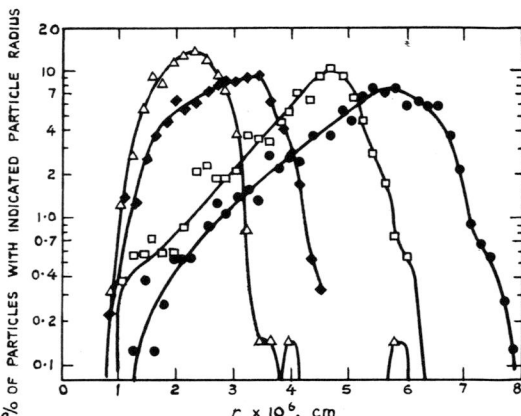

FIG. 9.5. Variation of the particle size distribution with the conversion of methyl methacrylate at 55 C. The experimental details are given in Table 9.8 [57].

9. Aspects of Emulsion Polymerization

V. ALTERNATE MODELS FOR PARTICLE FORMATION

A. The Harkins Hypotheses

The theory of Sect. 9.III is based on the Harkins hypotheses [65, 66] which are now restated. Particles are formed when monomer-swollen micelles absorb radicals. The surfactant on the surface of newly formed particles becomes by definition nonmicellar. Micellar soap is consumed by two mechanisms: when micelles are converted to particles, and when growing particles adsorb soap from the micelles. Particle formation stops when the micelles are used up. It is also assumed that the particle surfaces are kept saturated with soap during the Interval I of particle formation and that the total oil/water interfacial area remains constant in this period, equal to S cm^2/cm^3 of water.

The data of Sect. 9.IV show that the theory quite accurately predicts the results of numerous emulsion polymerization experiments but that there are discrepancies. It may be fruitful to reexamine the validity of the Harkins hypotheses.

The assumption that all the free surfactant is efficiently adsorbed onto unoccupied particle surfaces is not necessarily valid for all polymerizing systems. The available data directly measuring these effects are meager, but Table 9.5 shows that there can be substantial amounts of unadsorbed soap present even after particle nucleation is complete, as indicated by the constant value of $2r/c^{1/3}$. In such experimental situations the final particle number is likely to become much smaller than predicted by the theory (cf. Eq. 9.11).

Further deviation from the model may be caused if the rate of adsorption of surfactants onto the growing particle surfaces is slower than the growth rate of particle surfaces. Griskey [20] has recently shown that adsorption equilibria are reached in the order of 10 min. This is longer than the time needed for the completion of particle nucleation which, according to the theory [6, 7, 33], is of the order of 1 to 2 min.

Contrary to these data, there is evidence [37, 65, 67] that the micellar soap disappears at around the transition between Intervals

I and II. During Interval I the surface tension of the polymerizing latex was very low, corresponding to that of a micellar soap solution. At the conversion, when particle nucleation was proved or was expected to stop, the surface tension of the polymerizing latex was found to rise.

Evidently, more work is needed for directly elucidating the validity of the Harkins hypotheses. Most of the work to date tested the validity of these hypotheses indirectly, by testing the applicability of the theories based on these hypotheses as was done in Sect. 9.IV.

Several other questions arise. Even if the Harkins hypotheses were true, what are their theoretical validity limits? This question will be answered in Sect. 9.V.B. Furthermore, is it possible to form an alternate mechanistic picture which would lead to theoretical expressions similar to those based on the Harkins hypotheses? This question will be answered in Sect. 9.V.C.

B. *Particle Formation in the Absence of Micellar Surfactants*

It is well known that even in the absence of micellar soaps it is possible to obtain lattices having particle sizes in the 0.1 to 0.3 µm range. In some of the commercially used systems, the sulfate-ion end-groups of the polymer chains stabilize the particles; in other systems protective colloids perform the same function. Of course, stable lattices can also be obtained using submicellar concentrations of surfactants. A mechanism of particle nucleation for such systems is described below, based on the model proposed by Gardon [5]. Fitsch [68] obtained identical results via a different mathematical route.

The basic assumption is that particles are nucleated when radicals polymerize in the aqueous phase to some critical size and precipitate without being absorbed into existing particles.

These precipitated radicals subsequently absorb more monomer and grow. If an oligomeric radical collides with a particle before

9. Aspects of Emulsion Polymerization

it reaches its critical size for precipitation, it becomes trapped into this particle. As the particle population increases, radicals growing in the aqueous phase have a decreasing chance to precipitate and nucleate new particles.

A radical is assumed to reach the critical size for precipitation in a well-defined time, t_{prec}. During this time it moves, by Brownian motion, from its origin to an average distance L, which can be calculated from the mean diffusive displacement with the aid of the diffusion coefficient D.

$$L = (2Dt_{prec})^{0.5} \qquad (9.35)$$

If there is no particle within the distance L from the origin of a radical, this radical must precipitate and nucleate a new particle. On the other hand, if a radical originates within the distance L from the surface of some spherical oil-phase unit (micelle or particle), it may or may not collide with this unit. If the radical collides with the sphere, it becomes irreversibly absorbed into it. The probability of collision, $p(r,\lambda)$, is a function of sphere radius and of the distance, λ, from the site of radical origin to the sphere surface.

If the radical moves within the conelike space whose cross section is shaded in Fig. 9.6, it collides with the sphere; if it

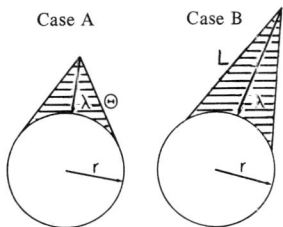

FIG. 9.6. Cross sections of geometrical models to calculate the probability that a radical formed at a distance λ from the particle will hit it. The particle radius is r, L is the mean free path, θ is the length of tangent from the point where the radical originates. In case A, $\lambda < \theta < L$; in case B, $\theta < L$, but $\lambda < L$. The shaded figure is the cross section of the volume which divided by the volume of a sphere either with radius θ in case A or with radius L in case B gives the probability [5].

moves outside the conlike space, it does not. Consequently, the value of $p(r,\lambda)$ equals the ratio of the volumes of the conelike space and that of a sphere with the radius θ or L, as defined in Fig. 9.6.

To calculate the rate of radical entry, dw/dt, into an isolated particle, a spherical shell at distance λ from the sphere surface is considered. The volume of this shell is $4\pi(r+\lambda)^2 \, d\lambda$. Since R is the number of radicals produced per unit volume of water per second, the number of radicals produced within the shell per second is $4\pi(r+\lambda)^2 R \, d\lambda$. The probability that any of these radicals will hit the particles is $p(r,\lambda)$, and Eq. (9.36) gives the total number of radicals entering into the particle per second.

$$\frac{dw}{dt} = 4\pi R \int_0^L p(r,\lambda) \, (r+\lambda)^2 \, d\lambda \tag{9.36}$$

The expression for $p(r,\lambda)$ is not shown here. The result of this integration is Eq. (9.37).

$$\frac{dw}{dt} = \pi r^2 LR = 0.25(\text{particle surface})LR \tag{9.37}$$

It follows from Eq. (9.37) that the diffusion of radicals into particles is not proportional to the particle radius, as it would be in homogeneous diffusion at spherical symmetry, but to the particle surface.

The total rate of adsorption of radicals by all particles present in 1 cm^3 of water is $\pi RL \Sigma n_i r_i^2$ and the new formula for particle nucleation is Eq. (9.38).

$$\frac{dN_t}{dt} = R(1 - \pi L \Sigma n_i r_i^2) \tag{9.38}$$

Accordingly, particle nucleation stops when the particles become so crowded that all radicals are adsorbed by them. This happens when $\pi L \Sigma n_i r_i^2$ exceeds unity.

The nucleation equation based on the Harkins model, Eq. (9.3), and the above derived Eq. (9.38) become mathematically identical if the parameters L and 4/S are interchanged. Consequently, the Harkins

9. Aspects of Emulsion Polymerization

model can only be valid if $S \geq 4/L$. Also, all other equations of Sect. 9.III can be derived from the present model by substituting for the parameter S the parameter $4/L$. The equivalence of these two parameters is dimensionally correct since the units of S are $cm^2/cm^3 = cm^{-1}$ and that of L is cm.

It is of some importance to estimate the order of magnitude of L. The value of t_{prec} can be calculated if the monomer concentration in water, the propagation constant, and the molecular weight of the freshly precipitated oligometric radical are known. Based on the data of Priest [69], Gardon [30] calculated a t_{prec} value of 0.07 sec for vinyl acetate. Assuming a plausible value for the diffusion coefficient, 10^{-6} cm^2/sec, Eq. (9.36) yields an L value of 3.7×10^{-4} cm. Similarly, Fitch and Tsai [68] estimated the value of L for methyl methacrylate to be 2.8×10^{-3} cm. If these numbers are correct, a "soapless" polymerization of these monomers would proceed as if the aqueous phase contained 0.074 or 0.011% sodium lauryl sulfate. At these surfactant concentrations the theoretical values of $4/S$ would become equal to the indicated values of L (Table 9.1). Consequently, this theory satisfactorily explains why surfactant-free or submicellar reaction systems give latices with particle sizes or particle numbers having orders of magnitude similar to those obtained with micellar surfactants.

For example, in soapless systems, with methyl methacrylate/water ratios in the 1/99 to 10/90 range, Fitch and Tsai [68] obtained 1.5×10^{13} particles per cubic centimeter of water using a redox initiator. Since the value of the parameter R is not known, exact quantitative comparison between theory and experiment is not possible. Nevertheless, it is of interest that data of Table 9.2 define for 0.08% sodium lauryl sulfate a particle number of 6.9×10^{13}. If K and R in the data of Fitch and Tsai were the same as in the data of Table 9.2, the theoretical value of N would be $(0.08/0.011)^{0.6}$ times less than that at 0.08% surfactant concentration, since 0.011% surfactant concentration corresponds to the L value of methyl methacrylate. Accordingly, 2.1×10^{13} particles per cubic centimeter of water would have been the "expected" particle number, quite close to the experimentally found 1.5×10^{13}.

C. Particle Nucleation at Submicellar Surfactant Concentrations

For the interpretation of experiments in which only the surfactant level is varied, two specific concentrations are important: the one (C_L) which corresponds to an S value equal to 4/L, and the critical micelle formation concentration, CMC. The Harkins model is applicable at soap concentrations exceeding the larger of these two specific values. The theory of Sects. 9.V.A and 9.V.B implies that, at lower concentrations, there would be no important effect of surfactant on the particle numbers and this is contrary to the experimental data.

For sodium lauryl sulfate the CMC is about 0.13% [11] and C_L is about 0.011% if the monomer is methyl methacrylate. Fitch and Tsai [68] reported particle numbers for methyl methacrylate polymerization with a redox initiator at about 10/90 water ratios. At surfactant concentrations of 0.0016%, 0.0096%, 0.03%, and 0.064% the particle numbers were 2.55×10^{13}, 3.08×10^{13}, 8.1×10^{13}, and 9×10^{13}, respectively. A similar trend of data was obtained by Roe [70].

According to the theoretical picture formulated by Roe [70], such data are inconsistent with the Harkins hypothesis of particle nucleation by absorption of radicals into monomer-swollen micelles. Roe believes that particles are formed, as in soapless particle nucleation, by the growing oligomeric radicals which precipitate out from water. The role of the surfactant is to stabilize these particles and inhibit their aggregation and coagulation. The surfactant present in the system absorbs onto the growing particle surfaces until the total surface area of the growing particles becomes equal to S, the surface that the soap molecules can occupy at the water/oil interface. When this happens, all the soap is depleted and no new particle can be stabilized. Radicals formed later precipitate into existing particles. This model, when mathematically formulated, leads to the same results as shown in Eqs. (9.11) and (9.12) with the parameters

9. *Aspects of Emulsion Polymerization* 241

also having the same significance; the only difference is that this model predicts a 1.48-times larger value for the particle number and a 1.13-times lower value for particle radius.

Roe's picture is intellectually satisfying because it explains why the particle number increases with surfactant concentration below CMC. One shortcoming of this model is that the experimental particle number was found to be proportional to a higher than the theoretical 0.6th power of soap concentration at concentrations below CMC [68, 70]. Also, this model overestimates the particle number and underestimates the particle size when compared to the experimental data obtained at soap concentrations above CMC and described in Sect. 9.IV.B.1.

D. *Critique of the Present Theories of Particle Nucleation*

From the standpoint of the interfacial aspects of emulsion polymerization it is important to understand the mechanism of particle formation and the processes that cause particle formation to stop at a low conversion.

The alternate hypotheses and the inherent uncertainties are summarized below:

1. Particles are assumed to form either when monomer-swollen micelles capture radicals from the aqueous phase or when oligomeric radicals precipitate from the aqueous phase. Either or both these assumptions may be valid. However the mathematical theory does not take into account reduction of particle number by possible aggregation of several particles.

2. The theory assumes that all primary radicals are used in nucleating particles and initiating or terminating chain growth. Nevertheless, it is not known whether the radical utilization efficiency is truly 100%.

3. At least two different processes were assumed to cause termination of particle formation. In soapless systems, this is caused by overcrowding of particles so that radicals growing in the aqueous phase become preferentially absorbed into existing particles

instead of precipitating to form new particles. For surfactant-containing systems, the mathematical theories require speedy and efficient adsorption of the available surfactant onto the growing particle surfaces so that particle nucleation stops when the surfactant-helping particle formation is used up. However, there are data available indicating that in some instances the efficiency of surfactant adsorption is low and the rate of adsorption is slow.

To complicate matters further, the independently determined parameters R, S, K, and L, which are used in the theory, are known only with limited precision. Also, at least some of the published experimental data giving particle sizes and particle numbers may be inaccurate.

Until a few years ago, the theory was tested only by checking whether the experimental particle numbers or Interval II rates were proportional to the theoretical 0.6th or 0.4th powers of soap or initiator concentrations. Recent theoretical developments allowed the use of more rigorous tests involving the absolute values of particle numbers, particle radii, and conversion rates. It is now evident that the present forms of the theories of particle formation describe reasonably well systems involving persulfate initiators, several anionic soaps, and styrene, methyl methacrylate, and butyl acrylate monomers. However, there is reason to suspect [10, 13, 14] that the framework of these theories does not fit oil-soluble initiators, nonionic surfactants, and monomers such as vinyl acetate, vinyl chloride, vinylidene chloride, and acrylonitrile. More experimental and theoretical work is needed for an understanding of the roles of such materials.

VI. SWELLING OF LATEX PARTICLES BY MONOMERS

The two key interfacial processes in emulsion polymerization are particle formation and swelling of the latex particles by unconverted monomers. The latter phenomenon will now be discussed in detail.

9. Aspects of Emulsion Polymerization

A. The Thermodynamics of Swelling

1. The Thermodynamics of Saturation Swelling

Even if the monomer is a good solvent for its polymer in bulk, it can swell latex particles only to a limited extent. While the free energy of mixing would favor the adsorption of unlimited amounts of monomer into the latex particles, this is compensated by the surface energy consumption as the particle size increases. In a finely divided system, as in lattices, the surface energy change on swelling is of the same order of magnitude as the free energy of mixing.

The theory concerns nonpolymerizing latex particles and monomers or solvents with limited solubility in water. It describes a three-phase system of latex particles saturated with monomer, an aqueous phase saturated with monomer, and a polymer-free monomer phase saturated with water. This theory was originally developed by Morton et al. [71] and was refined by Gardon [72].

The system is characterized by three parameters: V_m, the molar volume of the solvent; γ, the interfacial tension between the particles and water; and χ, the Flory-Huggins constant. It gives the dependence of the monomer volume fraction in the particles, $\emptyset_m$, and the radius of the swollen particle radius, r, on the unswollen particle radius, r_o. The relationship between these variables follows from their definition in Eq. (9.39):

$$\left(\frac{r}{r_o}\right)^3 = \frac{1}{1 - \emptyset_m} \tag{9.39}$$

The chemical potential difference, $\Delta\mu_s$, of spheres having finite and infinite radii is given by Eq. (9.40) due to Johnson [73]:

$$\Delta\mu_s = \mu(r) - \mu(\infty) = \frac{2V_m \gamma}{r} \tag{9.40}$$

The chemical potential of a solvent is reduced by $\Delta\mu_m$ when a polymer is dissolved in it. The Flory-Huggins theory [74] contains the gas constant, R^*, and the absolute temperature, T, and is described in Eq. (9.41):

$$\frac{\Delta\mu_m}{R^*T} = \ln \phi_m + (1 - \phi_m) + \chi(1 - \phi_m)^2 \quad (9.41)$$

Combination of these formulas so that the sum of $\Delta\mu_m$ and $\Delta\mu_s$ is set at zero gives Eq. (9.42), the Morton-Kaizerman-Altier equation:

$$\frac{-1}{1 - \phi_m} - \frac{\ln \phi_m}{(1 - \phi_m)^2} = \chi + \frac{2V_m \gamma}{R^*T} \frac{1}{(1 - \phi_m)^{5/3} r_o} \quad (9.42)$$

Figure 9.7 is a graphical presentation of Eq. (9.2) and shows that swelling increases with increasing radius, decreasing interfacial

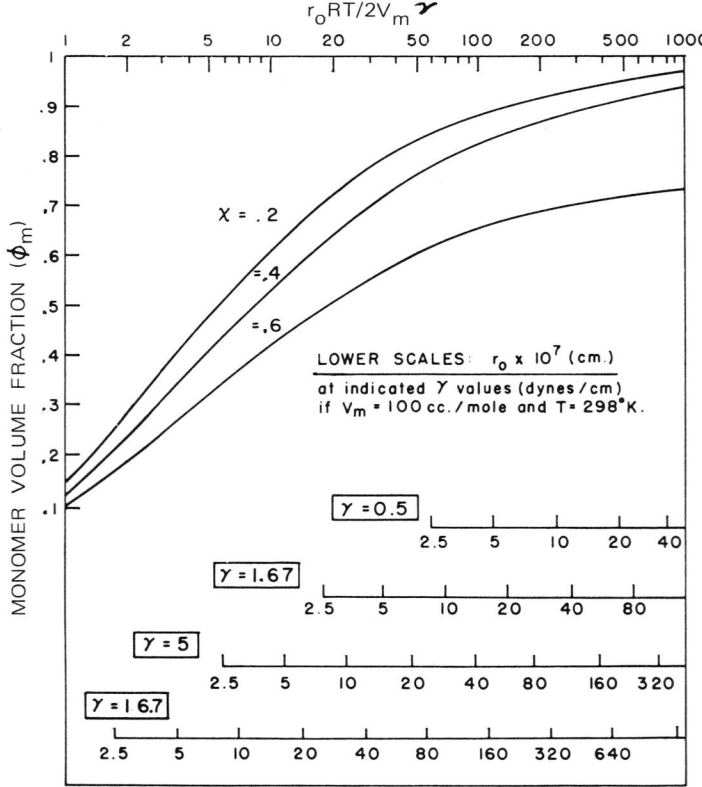

FIG. 9.7. Description of the saturation swelling of latexes according to Eq. (9.42). The unswollen radius is r_o, γ is the interfacial tension, χ is the Flory-Huggins constant, and V_m is the molar volume of the monomer [72].

9. Aspects of Emulsion Polymerization

tension, and improved solvent power as evidenced by lower values of χ. In the treatment of experimental data it is convenient to determine $\emptyset_m$ for a series of latices having various particle sizes, calculate the left-hand side and the square-bracketed term from $\emptyset_m$ and r_o, and determine χ and γ from the intercept and slope of a suitable plot. Such data are shown in series A of Table 9.10 and will be discussed later.

2. Partial Swelling

In partial swelling of latex particles the monomer is added at a level lower than needed for saturation so that there is no separate monomer phase present. If the actual vapor pressure of the system is p and the saturation vapor pressure is p_o, Eqs. (9.43) and (9.44) hold:

$$R^*T \frac{\ln p}{p_o} = \Delta\mu_m + \Delta\mu_s \tag{9.43}$$

$$\frac{-1}{1-\emptyset_m} + \frac{\ln(p/p_o) - \ln \emptyset_m}{(1-\emptyset_m)^2} = \chi + \frac{2V_m \gamma}{R^*T} \frac{1}{(1-\emptyset_m)^{5/3} r_o} \tag{9.44}$$

Equation (9.43) is suited for determining χ and γ with the same latex from the variation of $\emptyset_m$ with the partial pressure, as shown in series B of Table 9.10.

3. Values of Interfacial Tensions and Flory-Huggins Constants

In the experimental determinations of saturation swelling, an excess of monomer (or solvent) is stirred into a radical-free latex, and after centrifugation the monomer content of the lower, latex phase is determined. This value is corrected for the solubility of the monomer in water. Alternatively, monomer is added in various increments to the latex and its vapor pressure is recorded. Below saturation, the vapor pressure rises with monomer content: above saturation, the vapor pressure is constant. These are the so-called static methods of latex-swelling determinations. As will be shown in Sect. 9.VI.B, latex swelling can also be determined by "dynamic" methods involving emulsion polymerization experiments.

TABLE 9.10
Flory-Huggins Constants and Interfacial
Tensions Calculated from Latex Swelling Experiments

Series	Polymer	Swelling agent	Temp. (°C)	χ	γ (dyne/cm)	Ref.
A[a]	Polystyrene	Styrene	~25	0.43	4.5	71
		Toluene	~25	0.48	3.5	71
		Cyclohexanone	~25	0.53	2.1	71
	Poly(methyl methacrylate)	Methyl methacrylate	~25	0.58	4.9	72
			~25	0.55	1.6	72
B[b]	Polystyrene	Styrene	~25	0.35	35	72
		Styrene	30	0.30	31.2	75
		Styrene	60	0.26	12.4	75
	Poly(methyl methacrylate)	Methyl methacrylate	~25	0.55	11.0	72
	Poly(vinyl acetate)	Vinyl acetate	30	0.33	3.0	75
		Vinyl acetate	60	0.28	3.2	75
		Styrene	60	0.42	5.3	75
		Vinyl caproate	60	0.52	34.2	75
	Poly(vinyl caproate)	Vinyl caproate	60	0.10	9.5	75

[a] γ was assumed to be constant for various latexes having different r_o values. Saturation values of ϕ_m were established (cf. Eq. 9.42).

[b] ϕ_m of the same latex was determined at different partial saturations (cf. Eq. 9.44).

9. Aspects of Emulsion Polymerization

The available χ and γ values of Table 9.10 calculated according to the theory from static latex swelling data are plausible, though exact comparison with independently determined values is at present not possible.

Styrene and toluene are similar solvents and should give similar χ values polystyrene. The χ values obtained with toluene solutions of polystyrene vary between 0.3 and 0.44 [76-78] and are in the same range as the corresponding values in Table 9.10. The χ values of poly(methyl methacrylate) in Table 9.10 seem to be too high. Since this polymer dissolves in its monomer in bulk, $\chi < 0.5$ is expected [79]. It is possible that water dissolved in monomer raised the χ value by reducing the solvent power of the monomer.

Morton et al. [71] post-added soap to their polystyrene latex samples. In their swelling experiments the potassium laurate was at the critical micelle formation concentration and the monomer/latex interfacial tension was 14 dyne/cm, as determined by the DuNouy ring method. This value of γ was higher than the 4.5 dyne/cm shown in Table 9.10 because the ring method is not accurate when applied to soap solutions. The soap molecules adsorb into the interface rather slowly. When the interfacial film is suddenly ruptured in a ring measurement, the equilibrium conditions are not established. Roe and Brass [80] measured 4 dyne/cm interfacial tension between toluene and aqueous potassium laurate at the critical micelle concentration. These authors used an equilibrium method, the drop volume method, and their result is quite close to the value calculated from latex swelling data.

For lattices containing methyl methacrylate Gardon [72] obtained γ values of 4.9 and 11 dyne/cm by analyzing data obtained by saturation or partial swelling, respectively. The interfacial tension between a monomer-saturated poly(methyl methacrylate) latex and water-saturated monomer was found to be 11 dyne/cm when determined by the ring method.

B. Monomer Concentration in Latex Particles During Emulsion Polymerization

It is often possible to observe visually the sudden disappearance of monomer droplets during an emulsion polymerization experiment. The fractional conversion where this happens is exactly equal to the polymer weight fraction in the particles. From this value $\emptyset_m$ can be calculated and the result of such "dynamic" experiments can be compared to the "static" determinations.

Gardon [72] found by these two methods $\emptyset_m$ values of 0.75 and 0.74 for poly(methyl methacrylate) lattices containing 1% sodium lauryl sulfate soap on monomer with particle radii in the 0.12 to 0.14 µm range. All available "dynamic" $\emptyset_m$ values of poly(methyl methacrylate) and its monomer, covering a great variety of particle radii and surfactant levels, are in the range of 0.78 to 0.64 [45, 47, 72] while the "static" $\emptyset_m$ values are in the range of 0.62 to 0.77 [47, 72].

Similarly, Paoletti and Billmeyer [51] found identical $\emptyset_m$ values by "static" and "dynamic" determinations for polystyrene/styrene systems. The particle radius was 0.045 µm and 5.7% sodium lauryl sulfate was applied based on monomer. The $\emptyset_m$ value was 0.55. Gardon [72] found that 26 out of 28 experimental results taken from 13 references define "dynamic" and "static" $\emptyset_m$ values for this monomer/polymer system in the range of 0.48 to 0.75, for a great variety of experimental conditions. A literature survey of the results obtained with other monomers is presented by Gardon [72].

The identity of $\emptyset_m$ values obtained by "static" and "dynamic" experiments prove that during emulsion polymerization the latex particles are swollen by their monomers corresponding to the thermodynamic swelling equilibrium. In other words, diffusion of the monomers into the particles is much faster than their consumption by polymerization. Also, in practical recipes the monomer concentration in the particles is characteristic of each monomer/polymer system and varies relatively little with experimental conditions.

Since the swelling of latex particles during polymerization corresponds to a thermodynamic equilibrium, it is possible to check

9. *Aspects of Emulsion Polymerization*

the plausibility of the assumption made in Sect. 9.III that in Intervals I and II the monomer concentration is constant.

Freshly nucleated particles can be expected to have a radius of about 0.0025 μm radius, that of a micelle. The particles grow so that their radius increases about 50-fold during the polymerization. Figure 9.7 shows that the variation in the saturation concentration of monomer in the particles at a 50-fold change in radius will be relatively low if the interfacial tension is small and the monomer is a good solvent for the polymer so that the value of χ is low.

In many instances the variability of ϕ_m during Intervals I and II will be less than indicated in Fig. 9.7. In Interval II the particle surfaces become less and less completely covered by surfactant, causing the interfacial tension to rise and resulting in a possible reduction in the value of ϕ_m. This effect is compensated by a possible increase in the value of ϕ_m caused by progressively larger particle radii.

Direct determination of monomer concentration in particles indeed confirmed the expectation that the value of ϕ_m changed little with increasing conversion during Interval II of emulsion polymerization. The available Interval II data suggest a 5% increase [81] or 20% decrease [82, 83] in the value of ϕ_m for styrene and a random variation of 5% [72] for methyl methacrylate.

An unlikely high error of 25% in the value of ϕ_m used in calculating the key theoretical results for the Smith-Ewart rate, the particle number, or the particle radius by Eqs. (9.10), (9.11), (9.12), and (9.15) would cause errors in the predicted values of about 28%, 22%, or 6%, respectively. Evidently, the assumption of a constant value of ϕ_m for Intervals I and II of emulsion polymerization is a satisfactory approximation.

REFERENCES

1. M. V. Smith and R. H. Ewart, *J. Chem. Phys.*, 16:592 (1948).
2. R. N. Haward, *J. Polym. Sci.*, 4:273 (1949).
3. W. H. Stockmayer, *J. Polym. Sci.*, 24:314 (1957).
4. J. T. O'Toole, *J. Appl. Polym. Sci.*, 9:1291 (1965).

5. J. L. Gardon, *J. Polym. Sci.*, *Part A-1*, *6*:623 (1968)(Part I).
6. J. L. Gardon, *J. Polym. Sci.*, *Part A-1*, *6*:665 (1968)(Part III).
7. J. L. Gardon, *J. Polym. Sci.*, *Part A-1*, *9*:2763 (1971)(Part VII).
8. U. Ugelstad and P. C. Mark, *Brit. Polym. J.*, *2*:31 (1970).
9. A. E. Alexander and D. H. Napper, *Progress in Polymer Science* (A. D. Jenkins, ed.), Vol. 3, Pergamon, New York, 1971, p. 145.
10. J. W. Van der Hoff, *Vinyl Polymerization* (G. E. Ham, ed.), Vol. 1, Part 2, Dekker, New York, 1969, p. 1.
11. E. W. Duck, *Encyclopedia of Polymer Science and Technology* (H. F. Mark, N. G. Gaylord, and N. M. Bikales, eds.), Vol. 5, Interscience, New York, 1966, p. 801.
12. J. L. Gardon, (a) *Brit. Polym. J.*, *2*:1 (1970); (b) *Rubber Chem. Tech.*, *43*:74 (1970).
13. H. Gerrens, *Fortschr. Hochpolymer. Forsch.*, *1*:234 (1959).
14. B. M. E. Van der Hoff, *Advan. Chem. Ser.*, *34*:1 (1962).
15. F. A. Bovey, I. M. Kolthoff, A. I. Medalia, and E. J. Neehan, *Emulsion Polymerization*, Interscience, New York, 1955.
16. A. G. Parts, D. E. Moore, and J. G. Watterson, *Makromol. Chem.*, *89*:156 (1965).
17. J. G. Watterson, A. G. Parts, and D. E. Moore, *Makromol. Chem.*, *116*:1 (1968).
18. I. M. Kolthoff and I. K. Miller, *J. Amer. Chem. Soc.*, *73*:3055 (1951); ibid., *73*:5118 (1951).
19. E. H. Crooks, G. F. Trebbi, and D. B. Fordyce, *J. Phys. Chem.*, *68*:3592 (1964).
20. R. G. Grisbey, *Interface Conversion for Polymer Coatings* (P. Weiss and S. D. Cheever, eds.), Elsevier, New York, 1968, p. 338.
21. J. G. Brodnyan and G. L. Brown, *J. Colloid Sci.*, *15*:76 (1960).
22. W. M. Sawyer and S. H. Rehfeld, *J. Phys. Chem.*, *67*:1973 (1965).
23. Z. Pelzbauer, V. Hynkova, M. Bezdek, and F. Krabak, in *International Symposium on Macromolecular Chemistry*, Prague, 1965 (*J. Polym. Sci., Part C.*, *16*), O. Wichterle and B. Sedlacek, Chairmen, Interscience, New York, 1966, pp. 507-514.
24. S. H. Maron, M. E. Elder, and I. N. Ulevich, *J. Colloid Sci.*, *9*:89 (1954).
25. H. B. Klevens, *J. Colloid Sci.*, *2*:365 (1947).
26. M. Morton, P. P. Satatiello, and M. W. Altier, *J. Polym. Sci.*, *19*:457 (1956).

27. H. Gerrens, in *Polymer Handbook* (J. Brandrup and E. H. Immergut, eds.), Interscience, New York, 1966, pp. II-399 to II-419.
28. M. S. Matheson, E. E. Auer, B. B. Bevilacqua, and E. J. Hart, *J. Amer. Chem. Soc.*, 73:1700 (1951).
29. M. S. Matheson, E. E. Auer, B. B. Bevilacqua, and E. J. Hart, *J. Amer. Chem. Soc.*, 71:497 (1949).
30. J. L. Gardon, *J. Polym. Sci.*, Part A-1, 6:2853 (1968)(Part V).
31. G. V. Schulz, *Z. Phys. Chem.* (Frankfurt) 8:290 (1956).
32. K. E. Barrett and H. R. Thomas, *J. Polym. Sci.*, Part A-1, 7:2621 (1969).
33. J. L. Gardon, *J. Polym. Sci.*, Part A-1, 6:643 (1968)(Part II).
34. W. V. Smith, *J. Amer. Chem. Soc.*, 71:4077 (1949).
35. G. M. Burnett and R. S. Lehrle, *Proc. Roy. Soc.*, Ser. A, 253:331 (1959).
36. W. V. Smith, *J. Amer. Chem. Soc.*, 70:3695 (1948).
37. E. Barthalome, H. Gerrens, R. Herbeck, and H. M. Weitz, *Z. Elektrochem.*, 60:334 (1956).
38. B. M. E. Van der Hoff, *J. Polym. Sci.*, 44:241 (1960).
39. I. D. Robb, *J. Polym. Sci.*, Part A-1, 7:417 (1969).
40. M. Hamada, M. Nomura, H. Kohima, W. Eguchi, and S. Nagata, *J. Appl. Polym. Sci.*, 16:881 (1972).
41. H. Gerrens, *Ber. Bunsenges. Physik. Chem.*, 67:47 (1962).
42. C. E. M. Morris, A. E. Alexander and A. G. Parks, *J. Polym. Sci.*, Part A-1, 4:985 (1966).
43. D. M. French, *J. Polym. Sci.*, 32:396 (1958).
44. H. Gerrens, W. Fink, and E. Hohnlein, in *Macromolecular Chemistry*, Prague, 1965 (*J. Polym. Sci.*, Part C, 16), O. Wichterle and B. Sedlacek, Chairmen, Interscience, New York, 1967, p. 2781.
45. D. Hummel, G. Ley, and C. Schneider, *Advan. Chem. Ser.*, 34:60 (1962).
46. H. Gerrens and E. Kohnlein, *Z. Elektrochem.*, 64:1199 (1960).
47. Z. Manyasek and A. Rezabek, *J. Polym. Sci.*, 56:47 (1962).
48. G. M. Burnett, R. S. Lehrle, D. W. Overnall, and F. W. Peaker, *J. Polym. Sci.*, 29:417 (1958).
49. R. Pastiga, M. Litt, and V. Stannett, *J. Phys. Chem.*, 64:801 (1960).
50. M. Morton, P. P. Salatiello, and H. Landfield, *J. Polym. Sci.*, 8:279 (1953).

51. K. P. Paoletti and F. W. Billmeyer, *J. Polym. Sci., Part A-1*, *1*:2679 (1963).
52. J. W. Brietenbach, K. Kuchner, H. Fritze, and H. Tarnowiecki, *Brit. Polym. J.*, *2*:13 (1970).
53. W. S. Zimmt, *J. Appl. Polym. Sci.*, *1*:323 (1959).
54. T. F. Wall, R. W. Powers, G. D. Sachs, and G. S. Tent, *J. Amer. Chem. Soc.*, *69*:904 (1947).
55. D. H. Napper and A. G. Parts, *J. Polym. Sci.*, *25*:113 (1962).
56. P. M. Hay, J. C. Light, L. Marker, R. W. Murray, A. T. Santonicola, O. J. Sweeting, and J. C. Wepsic, *J. Appl. Polym. Sci.*, *5*:23 (1961).
57. J. L. Gardon, *J. Polym. Sci., Part A-1*, *6*:687 (1968)(Part IV).
58. G. J. Ley, Ch. Schneider, and D. O. Hummel, *Amer. Chem. Soc., Polymer Preprints*, *7*(2):725 (1966).
59a. S. W. Benson and A. M. North, *J. Amer. Chem. Soc.*, *81*:1339 (1959).
59b. A. M. North and G. A. Reed, *J. Polym. Sci., A*, *1*:1311 (1963).
60. B. M. E. Van der Hoff, *J. Polym. Sci.*, *33*:487 (1958).
61. J. W. Van der Hoff, E. B. Bradford, H. L. Tarkowsky, and W. B. Wilkinson, *J. Polym. Sci.*, *50*:265 (1961).
62. S. Jovanovic, J. Romatowski, and G. V. Schulz, *Makromol. Chem.*, *85*:187 (1965).
63. J. C. Bevington, H. W. Melville, and R. P. Taylor, *J. Polym. Sci.*, *12*:449 (1954); ibid., *14*:463 (1954).
64. D. C. Sundberg and J. D. Eliassen, *Polymer Colloids* (R. M. Fitch, ed.), Plenum, New York, 1971, p. 153.
65. W. D. Harkins, *J. Amer. Chem. Soc.*, *69*:1428 (1947).
66. W. D. Harkins, *J. Polym. Sci.*, *5*:217 (1950).
67. V. I. Yeliseyeva, P. I. Zubov, and V. F. Malofeyevskaya, *Vysokomol. Soedin.*, *7*:1348 (1965).
68. R. M. Fitch and C. H. Tsai, *Polymer Colloids* (R. M. Fitch, ed.), Plenum, New York, 1971, p. 73.
69. W. J. Priest, *J. Phys. Chem.*, *56*:1077 (1952).
70. C. P. Roe, *Ind. Eng. Chem.*, *60*:20 (1968).
71. M. Morton, S. Kaizerman, and M. W. Altier, *J. Colloid Sci.*, *9*:300 (1954).
72. J. L. Gardon, *J. Polym. Sci., Part A-1*, *6*:2859 (1968)(Part VI).
73. C. A. Johnson, *Surface Sci.*, *3*:429 (1965).
74. P. J. Flory, *Principles of Polymer Chemistry*, Cornell, Ithaca, 1953, pp. 511-518, 576-581.

9. *Aspects of Emulsion Polymerization*

75. E. Vanzo, R. H. Marchessault, and V. Stannett, *J. Colloid Sci.*, *20*:62 (1965).
76. G. M. Bristow and F. M. Watson, *Trans. Faraday Soc.*, *54*:1742 (1958).
77. M. L. Huggins, *Ann. N. Y. Acad. Sci.*, *44*:431 (1943).
78. C. E. H. Bawn, R. F. S. Freeman, and A. R. Kamaliddin, *Trans. Faraday Soc.*, *46*:677 (1950).
79. J. L. Gardon, in *Encyclopedia of Polymer Science and Technology* (H. F. Mark, N. G. Gaylord, and N. Bikales, eds.), Vol. III, Interscience, New York, 1966, pp. 833-862.
80. C. P. Roe and P. D. Brass, *J. Colloid Sci.*, *9*:602 (1954).
81. M. R. Grancio and D. J. Williams, *J. Polym. Sci., Part A-1*, *8*:2617 (1970).
82. B. M. E. Van der Hoff, *J. Phys. Chem.*, *60*:1250 (1956).
83. S. H. Herzfeld, A. Roginsky, M. L. Corrin, and W. D. Harkins, *J. Polym. Sci.*, *5*:207 (1950).

10.

BIOLOGICAL PHENOMENA AND INTERFACES

David Allan Cadenhead
Department of Chemistry
State University of New York
at Buffalo
Buffalo, New York

Robert E. Baier
Department of Biophysics
Roswell Park Division
State University of New York
at Buffalo
Buffalo, New York

I.	INTRODUCTION	255
II.	BIOLOGICAL MEMBRANES	257
	A. Membrane Models	259
	B. New Techniques and Some Findings	264
	C. Outstanding Problems	267
III.	BIOLOGICAL POLYMERS AND INTERFACES	269
	A. Biomedical Implants	270
	B. Surfaces of Collagen and Keratin	271
	C. Surgical and Dental Adhesives	273
	D. Maritime Interfaces	273
IV.	SUMMARY AND PROSPECTS	274
	REFERENCES	276

I. INTRODUCTION

Recent books describing specialized situations where biological interfacial phenomena dominate include the volumes edited by Maibach and Rovee on *Epidermal Wound Healing* [1], by Manly on *Adhesion in Biological Systems* [2], by Dale on *Management of Arterial Occlusive Disease* [3], by Blank on *Surface Chemistry of Biological Systems* [4], by Lasslo and Quintana on *Surface Chemistry and Dental Integuments*

[5], and by Brown on *Chemistry of the Cell Interface* [6]. All emphasize the present view that most biological reactions are, in fact, surface chemical reactions in which one or all of the reactants are spatially restricted. The many authors contributing to the referenced volumes describe problems such as the healing of lacerations and surgical incisions, the use of artificial materials in the human vascular tree (as synthetic blood conduits), the adverse events of biological fouling of structures in the marine environment, and the spontaneous accumulation of organic films on inorganic surfaces in the human oral cavity.

In another recent volume, Eirich systematized current thoughts on "interface conversion" in nonbiological circumstances [7]. It is now generally accepted that similar conversions must also take place between cell surfaces and their substrates prior to bioadhesive events or the accumulation of gross biological deposits such as thrombus, sclerotic plaques, dental plaques, and algae slimes [8].

Most chemists are familiar with the ubiquitous use of biological polymers such as gelatin and casein in commercial applications, but not so familiar that they consider the enormous dependence on the artful achievement of proper surface properties of such protein preparations for their efficacy in most technical applications. For example, the art and science of producing gelatin of an acceptable quality for photographic uses requires continuing study and improvement [9]. The discussion in the following paragraphs of a short series of biological interfacial situations should make it abundantly clear that a common interfacial chemistry exists among numerous apparently diverse biochemical fields, and in each of these fields the skills of chemists familiar with interfacial synthesis and interface conversion techniques will be of great benefit. The purpose of this short chapter is to provide, first, an introductory review of biomembrane organization, and, second, the briefest possible overview of the many other areas where biological polymers and interfaces are of key significance, as a stimulus to physical scientists who might wish to contribute to progress in these areas.

10. *Biological Phenomena and Interfaces*

II. BIOLOGICAL MEMBRANES

As described in Chap. 6, specific reactions do occur at and within even two-dimensional monomolecular films. These reactions may typify events at or within the membranes that surround and abundantly populate living cells. Each living cell is surrounded by membranous material and many are also filled with membranous two-dimensional "floor space" upon which most interesting and important biochemical syntheses occur.

When thinking of biochemical phenomena at interfaces, it becomes difficult, if not impossible, to avoid considering cell membranes. Certainly, it is true that biological membranes play an important role in almost all cellular phenomena. In any discussion of biochemical interfaces, a treatment of cell membrane structure and function is a must, though to the uninitiated it might seem remarkable that what for many years was simply considered a permeability barrier is now regarded as a dynamic, multifunctional structure. In a real sense, membranes have replaced nucleic acids as the prime area of intensive biochemical research. Thus, in the past year alone some ten thousand publications have appeared, necessitating the appearance of an abstracting service devoted to cell membranes [10]. While reviews of cell membrane structure and function have frequently been made [11, 12, 13] these have generally been directed toward practicing membranologists. Here, we provide only a brief review for the uninitiated but interested reader.

Two sets of circumstances have greatly assisted this increased interest in cell membranes. Firstly, rapid advances have been made in the isolation and purification of cell membrane components (lipids, proteins, and carbohydrates). This is particularly true for lipids where a combination of gas and thin-layer chromatography had led to an accurate knowledge of the lipid composition within a membrane and even of the hydrocarbon chain length and degree of unsaturation within a given lipid species. At this time, increasing numbers of synthetic phospholipids are becoming available to researchers and have proved particularly useful in many model membrane studies. Secondly, the application of a wide variety of investigational tools, many of which have been developed or adapted specifically for membrane investigations, has been made to both model membrane systems and real cell membranes.

In seeking an understanding of the behavior of cell membranes, there has been a tendency to attack the problems of structure and function separately; however, the results have shown that the two are closely related. Certainly, we cannot relegate the structural role totally to lipids or glycolipids (a combination of carbohydrates, usually complex saccharides, and lipids; see Fig. 10.1) or the functional role exclusively to proteins or glycoproteins (a combination of carbohydrates and proteins). Frequently, reference is made to lipoproteins (a combination of lipids and proteins). Such entities would seem to provide the necessary bridge between structure and function; but since the extent, strength, and specificity of the binding between the lipid and protein areas within given molecular clusters are often not well understood, we have only succeeded in redefining the problem.

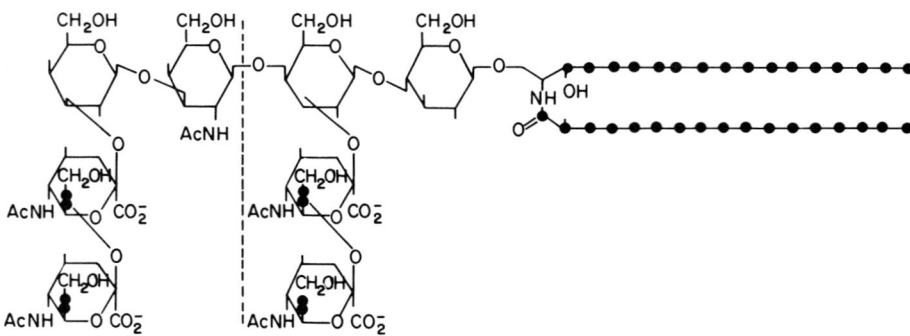

FIG. 10.1. A typical glycolipid, in this case a ganglioside, isolated from embrionic human brain and shown to be capable of influencing the functional capability of some membranes. (Ac = Acetyl, ● = hydrogen saturated carbon atom).

A. Membrane Models

The first important proposal relating to membrane structure was made by Gorter and Grendel, who suggested the bilipid layer as a permeability barrier [14]. In this arrangement, the polar heads were oriented outward toward each aqueous phase with the hydrocarbon chains inward toward the lipoid or oil phase. This schematic was incorporated in the Danielli-Davson model where protein was envisaged as attached, in globular form, to the exterior of the lipid bilayer [15] through hydrophilic interactions (see Fig. 10.2). Later this was modified to incorporate an initial denatured layer of protein against the bilayer with a second layer of globular protein. Protein penetration of the bilayer was also included to create hydrophilic pores to help explain the ability of membranes to transport ions [16]. (See Figs. 10.3 and 10.4.) Initially the question of whether the oriented hydrocarbon chains from each side of the bilayer left an isotropic lipoid region in the center, or met or interdigitated,

EXTERIOR

LIPOID

INTERIOR

FIG. 10.2. The Danielli-Davson model membrane [15]. The model incorporates the Gorter-Grendel bilayer [14] with spherical globular protein attached through hydrophilic interactions and effectively shielding the polar lipid regions. (Reproduced by permission of the Wister Institute of Anatomy and Biology, Philadelphia, Pennsylvania.)

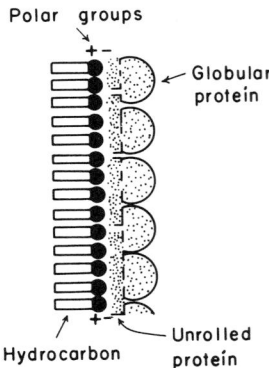

FIG. 10.3. A modified version of the original Danielli-Davson membrane [16] suggesting the existence of two protein layers, one unrolled or spread immediately adjacent to the polar lipid region, the second of globular protein. Only half of the membrane is illustrated. (Reproduced by permission of the Cold Spring Harbor Laboratory, Cold Spring Harbor, New York.)

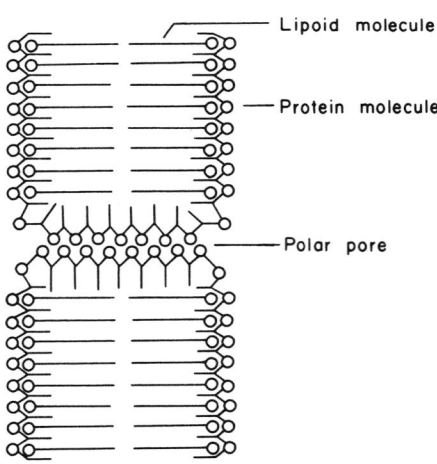

Diagram of pore of membrane

FIG. 10.4. A refinement of the original Danielli-Davson membrane model by Stein and Danielli [16] to help explain the observed membrane permeabilities of various ionic species. Denatured or unrolled protein is envisaged as penetrating the lipid bilayer. (Reproduced by permission of the Aberdeen University Press, Aberdeen, Scotland.)

10. Biological Phenomena and Interfaces

was left open. Subsequently, however, Danielli [17] has shown that the chains must meet, in order to minimize surface free-energy between anisotropic and isotropic regions. However, bearing in mind the variation in hydrocarbon chain length, we can conclude that some interdigitation must occur. Robertson [18], on the basis of electron microscopy and X-ray diffraction data, modified this concept by proposing that none of the protein was globular but, rather, was entirely stretched out across the exterior of the bilipid layer, to create the so-called unit membrane. Moreover, since alternating light and dark layers, forming approximately 80Å-thick sandwich-like patterns, were observed with samples prepared from a wide variety of membranes, it was further postulated that such structure existed in all membranes (see Fig. 10.5). This was in spite of the fact that analyses reveal vast differences between myelin with only 25% protein and mitochondrial membrane containing as much as 75% protein.

In the early 1960s the increasing flow of analytical information from a number of investigational techniques led to a questioning of the extrapolation of the lipid bilayer model from myelin to other more highly functional membranes. The probing became so extensive that the very existence of the lipid bilayer as a permeability barrier and structural unit was called into question, except possibly for the insulating myelin sheath where the X-ray diffraction data seemed incontravertial. In its place was proposed a structural protein [19], which replaced the continuous bilayer with a sequence of protein, lipid, or lipoprotein building blocks. While a structural role for protein or glycoprotein was not unreasonable in view of the discovery of nonenzymatic membrane proteins, the probability of finding a "*common* structural protein" in all membranes is now more or less rejected [20]. Indeed, the end result of this somewhat tumultuous decade has been the reinstatement of the lipid bilayer as an important constituent of almost all membranes. The difference between the situation now and that which existed in the 1950s is that the bilipid layer now rests on a firm experimental foundation. However, thinking has by no means reverted to the simple membrane models of the 1930-1960 era, and many refinements have been made.

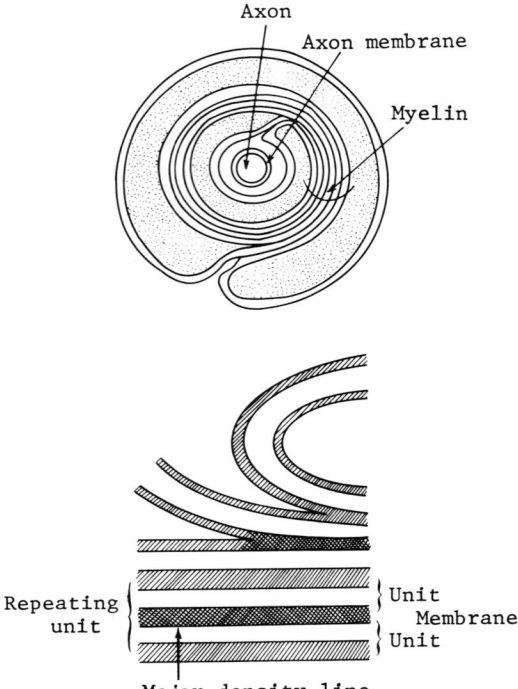

FIG. 10.5. The nature and formation of a myelinated axon nerve fibre. The illustration at the top depicts a transverse section of such a fibre, the myelin being created from the outer membrane of an oligodendrocyte cell. Partial nonsymmetry in the external and cytoplasmic membrane surfaces coupled with the mode of formation of the myelin can result in the repeating unit (as seen by X-ray diffraction) differing from the "unit membrane" (bottom illustration).

The membrane is now envisaged as a highly dynamic structure some 70-80 Å thick having a lipid bilayer core. Proteins, primarily globular, are thought to "float" in this sea of lipid, some essentially locating at the polar interface, others penetrating the lipid core (see Figs. 10.6 and 10.7). The relative amounts of lipid and protein will, of course, show considerable variation from one membrane to another. Carbohydrates associated with the membrane are thought always to be associated with either lipid to form glycolipids or with protein to form glycoproteins. Even the lipid bilayer is regarded as capable of segregation into distinct phases when some of the wide variety of lipid components become immiscible [21]. (See Table 10.1 and Fig. 10.8.)

FIG. 10.6. A cross-sectional representation of the cell membrane illustrating a lipid-protein mosaic. In addition to the concept of fluidity [12] the model depicts mainly globular protein, some of which completely penetrates the bilipid layer, and indicates that a substantial portion of the polar lipids are exposed (compare with the various Danielli models, Figs. 10.2 through 10.4). (Reproduced by permission of the American Association for the Advancement of Science.)

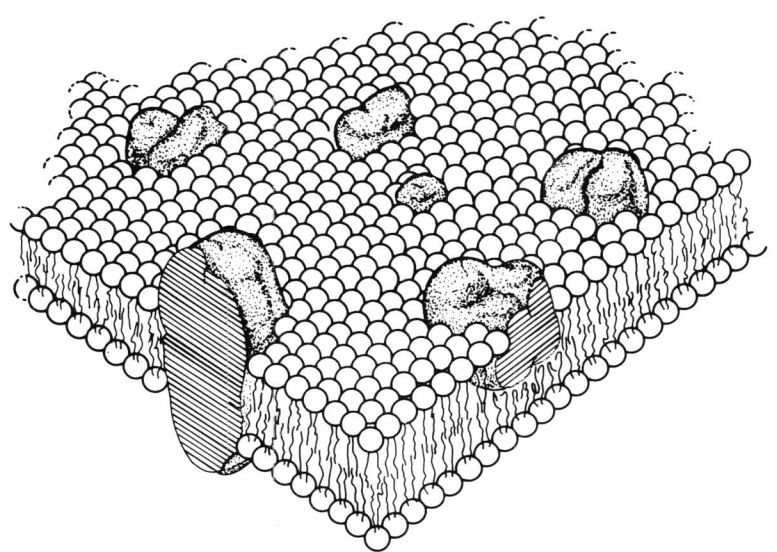

FIG. 10.7. A three-dimensional representation of Fig. 10.6 depicting protein "afloat in a sea of lipid." Since component mobility is relative the proteins possess the ability to segregate or to remain isolated. (Reproduced with permission from *Membrane Structure and Function* (L. I. Rothfield, ed.), Academic Press, London and New York, 1972.)

TABLE 10.1

Cell Membrane Lipids (Molar % Total Lipids)[a]

Lipid	Human CNS myelin	Bovine PNS myelin	Human erythrocyte	Rat liver mitochondria
Phosphatidylethanolamine	13	13	14	27
Phosphatidylcholine	11	10	15	43
Phosphatidylserine	5	7	5	trace
Phosphatidylinisotol	2	2	4	7
Diphosphatidylglycerol	trace	trace	4	9
Cholesterol	40	39	42	6
Sphingomyelin	4.6	13.0	9.4	4.4
Cerebroside	15.4	11.5	trace	

[a]Values selected from Table 1 of review of J. S. O'Brien, J. Theor. Biol., 15:307 (1967).

B. New Techniques and Some Findings

The reestablishment of the bilipid layer can be credited to a number of different techniques, two of which provide the strongest evidence: X-ray analysis and differential scanning calorimetry. Some workers, having studied the highly ordered membranes of myelin and retinal rods using standard diffraction techniques, have expanded their studies to a variety of other membranes using aqueous dispersions of membrane fragments to obtain the diffraction data [22]. Comparison with aqueous dispersions of the extracted lipids verified the existence of the lipid bilayer. In some cases, it would appear that the bilayer is the major structural component. Thus, Steim et al., using differential scanning calorimetry to investigate the membranes and extracted lipids of Mycoplasma laidlawii [23], have shown that, after growing such organisms on a selected lipid diet, excellent correspondence can be obtained between the thermal transition of lipids in their membranes and in the aqueous dispersions of the extracted lipids.

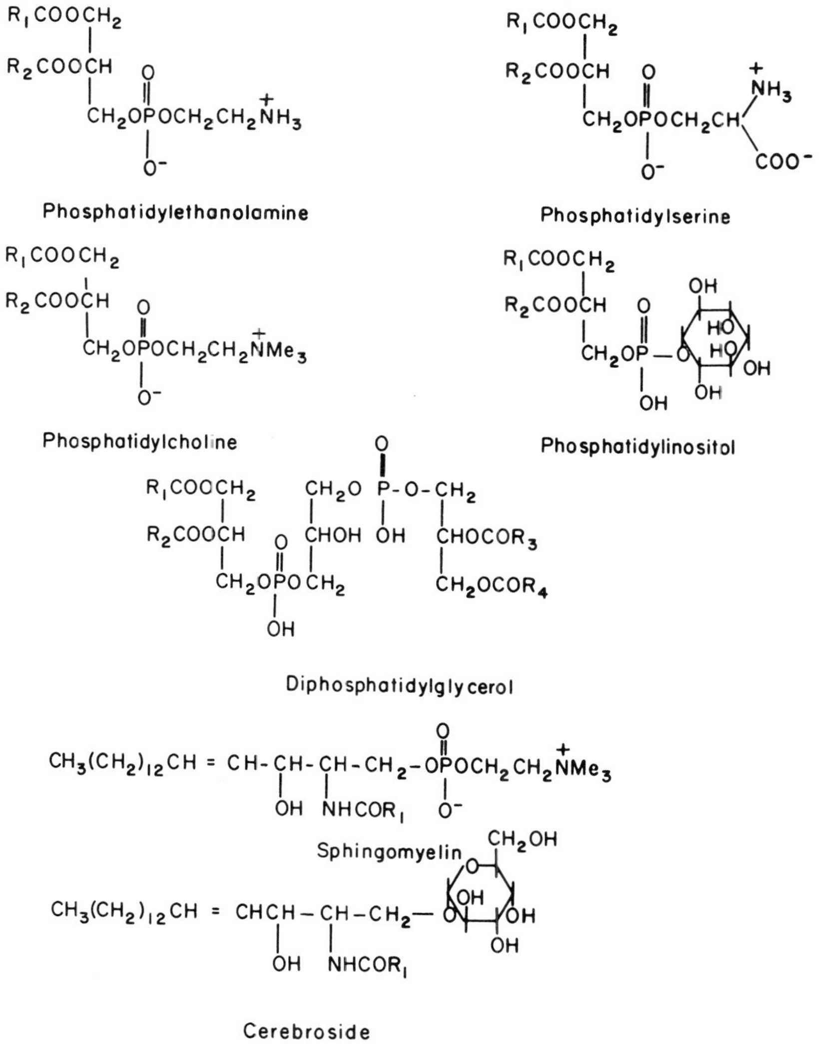

10. Biological Phenomena and Interfaces

R_1COOCH_2
R_2COOCH
$CH_2OPOCH_2CH_2\overset{+}{N}H_3$ (with P=O and O⁻)

Phosphatidylethanolamine

R_1COOCH_2
R_2COOCH
$CH_2OPOCH_2CH\overset{\diagdown}{\underset{\diagup}{}}\overset{+}{N}H_3 / COO^-$ (with P=O and O⁻)

Phosphatidylserine

R_1COOCH_2
R_2COOCH
$CH_2OPOCH_2CH_2\overset{+}{N}Me_3$ (with P=O and O⁻)

Phosphatidylcholine

R_1COOCH_2
R_2COOCH
CH_2OP—(inositol ring with OH groups)

Phosphatidylinositol

$R_1COOCH_2 \quad CH_2O\overset{O}{\overset{\|}{P}}-O-CH_2$
$R_2COOCH \quad CHOH \; OH \quad CHOCOR_3$
$CH_2OPOCH_2 \qquad\qquad CH_2OCOR_4$
$\quad OH$

Diphosphatidylglycerol

$CH_3(CH_2)_{12}CH = CH-CH-CH-CH_2-O\overset{O}{\overset{\|}{P}}OCH_2CH_2\overset{+}{N}Me_3$
$\qquad\qquad\qquad OH \;\; NHCOR_1 \;\; O^-$

Sphingomyelin

$CH_3(CH_2)_{12}CH = CHCH-CH-CH_2-$ (sugar ring with OH, CH$_2$OH)
$\qquad\qquad\qquad OH \;\; NHCOR_1$

Cerebroside

FIG. 10.8. Molecular structures of some typical membrane lipids (see Table 10.1). R_1 through R_4 depict hydrocarbon chains, typically C_{12} through C_{20} with one to four double bonds typically located in the chain attached at the 2-glycerol position. The only important membrane lipid omitted is cholesterol (see Table 10.1). No other steroid molecules occur to a significant degree in normal cell membranes.

The dynamic or fluid membrane concept has been proselytized by Singer and Nicolson in their recent review [12] where considerable experimental support is listed. Other favorable experimental evidence is provided in the proceedings of some recent cell membrane conferences [24, 25, 79]. Perhaps the most exciting approach to demonstrate cell surface fluidity has been the use of spin-label probes which were incorporated in sonicated liposomes (particles of about 300 Å diameter consisting of a near-spherical single bimolecular layer of lipid) and in membranes. Initially, such work was carried out using molecules bearing little resemblance to cell-membrane lipid components. Led by McConnell and co-workers [26, 27] this criticism has been answered, and spin-labeled components presently in use consist of modified membrane lipids in which the free radical group (usually a dimethyl substituted oxazolidine nitroxide ring), either is substituted for the polar group of a fatty acid, of a phosphatidyl choline (lecithin), or of cholesterol, or is attached to a hydrocarbon chain. The results clearly demonstrate the increased fluidity of the hydrocarbon chains proceeding from the polar region toward the interior of the bilayer. They also provide a measure of lateral movement of lipids as a whole within the membrane. One result suggests that a given molecule will exchange position with its nearest neighbor several thousand times per second [28]. The precise figures given in quantitative statements may be called into question since the spin-label group constitutes a perturbation of the immediate environment [29, 30]; however, cross correlation with other techniques suggests that the findings concerning fluidity are at least of the correct order of magnitude.

The ability of proteins to retain secondary, and perhaps tertiary, structure at the polar interface has been frequently demonstrated both with actual membranes and with model membrane systems. Studies of membrane protein using optical rotary dispersion and circular dichroism techniques have clearly demonstrated the retention of α-helical structure [31]. Partial stretching or denaturation may occur in other cases. Confirmatory evidence has been obtained through air-water interface monolayer studies [32, 33], and with liposomes [34]. (See Ref. 35 for a description of monolayer techniques.)

The ability of proteins (or glycoproteins) to penetrate the bilayer appears to have been successfully demonstrated using a new variation of the electron microscopy theme called "freeze fracturing." In order to avoid structural rearrangement during dehydration of a sample, an alternate procedure of freezing and fracturing the membrane and subsequently subliming away the resultant ice has been developed. By labeling the internal and external polar membrane regions, it has been shown that membrane fracture usually occurs down the center of the lipid bilayer. Because of this, it has been possible to observe indentations of approximately 70-75 Å in diameter, sometimes on one side or the other but occasionally in corresponding positions on both sides of the membrane, indicating complete penetration of the bilayer presumably by a protein or glycoprotein [36]. Confirmation that such penetrant particles are probably proteins or glycoproteins has been provided by Bretscher [37], who used a labeling agent which binds to proteins but which cannot penetrate the membrane and, therefore, labels only on the outside. The number of particles found shows good correspondence with that obtained for penetrating particles by freeze-etching.

C. *Outstanding Problems*

Having clearly determined a reasonable general structure for cell membranes, investigators are now concerning themselves with the finer details for it is here that membrane function and reactivity are involved. The question of ion permeability has been simplified through the use of sonicated liposome model membranes (spherical particles consisting of a continuous lipid bilayer). Beyond all doubt, the permeability of these purely lipid particles is several orders of magnitude too low to explain that of a living cell membrane [38]. A similar statement may be made concerning the ion selectivity [39] of pure lipid liposomes. Thus the ion permeability and selectivity of membranes may be ascribed to proteins or lipoproteins but not to pure lipid regions of the membrane. Similarly, the question of hormone or drug-receptor interactions is also dependent on proteins or lipoproteins but not on pure lipids [40].

The question of membrane protein conformation remains a difficult problem. Nevertheless, the effective "on-off" control of membrane function achieved by gel-liquid crystalline membrane lipid transitions [41, 42, 43, 44, 45, 46] is highly suggestive that an interfacial environment may be necessary for functionality. Certainly, polymeric molecules have been shown to achieve desirable conformations at an interface which have not been attained in a three-dimensional environment [47]. One other fascinating possibility is that suggested by Sumper and Trauble [48], that locating an enzyme in a membrane will improve its efficiency through an enhanced number of receptor-substrate interactions. The lipid bilayer is envisaged as a temporary trap or storage depository for a substrate molecule, the final encounter with the enzyme being achieved through lateral diffusion [50].

Although we may have acceptance of a qualitative model of a general cell membrane and some progress has been made toward making this model more quantitative, a great deal remains to be done. Much of this progress was achieved using the freeze-fracturing technique already described and by the labeling studies of Bretscher [36, 37, 49]. Another interesting approach has been described by Trauble and Overath [50] who have attempted to identify the fraction of membrane lipids involved in a phase transition (and not associated with protein) as seen (1) from the interior of the hydrophobic lipid region and (2) from the membrane-aqueous interface. Trauble obtained his data through the use of fluorescent membrane probes which were assumed to locate exclusively in one or the other region. Based on such information, it was possible to establish values for the fraction of lipid associated with protein through hydrophobic and through hydrophilic interaction.

Further insight may be gained into the nature of the cell membrane when it is correctly regarded as two, not necessarily identical,

interfaces, a concept similar to that of Langmuir's "duplex" film
[51]. In almost all of the work already done, little regard has
been paid to the possibility of an asymmetric membrane model, yet
there exists a considerable amount of evidence that this is the
case. Bretscher, in his recent review [49], has strongly advocated
this concept; while occasionally his views might be labeled speculative, they are perhaps justified by the importance of the topic and
the stimulation that they should provide to membrane research. At
this time, all we have established is a rather fuzzy qualitative
picture of a "general" cell membrane. As we focus on the large
number of existing membranes and on both external and internal surfaces, we can expect to see a large number of variations of our
original structural concept.

III. BIOLOGICAL POLYMERS AND INTERFACES

Even outside the realm of biological membranes, biological polymers
at interfaces provide significant new conditions for biochemical
reactions. Biologically active proteins of a variety of types have
been attached to surfaces of inorganic materials and of synthetic
polymers. Enzymes bound to these already-attached protein layers
might ultimately be used for treatment of certain diseases, including
leukemia and enzyme deficiencies. They have already been proposed
for use in chronic dialysis of patients having kidney deficiencies.
Surface-attached enzymes are also being used, with great economic
return to the device manufacturers, in clinical chemical analyses.
These enzymes are entrapped at surfaces, within gels, or inside membranes surrounding electrodes. Examples of the latter devices are
certain ion-specific electrodes. Other examples of biologically
active materials which can be attached to synthetic substrates
include the antibodies and antigens which can, in turn, be used to
selectively adsorb constituents from blood or tissue fluid. The
methods of preparation of these several attached proteins need not
be novel. Interfacial synthetic chemists will recognize merit in
direct extrapolations of now well-worn laboratory methods. It might

not be equally apparent, however, from discussion of synthetic surfaces modified by the application of biopolymer films, that beneficial modifications might also be made to natural biological surfaces. Interfacial synthetic schemes are also needed whereby the biochemically adverse reactivity of certain protein surfaces, correlating with their lack of immunological acceptance in implants, for example, might be effectively masked.

A. *Biomedical Implants*

Adsorption of proteins from blood onto prosthetic materials has been recognized as the first event that occurs in the complex chain of events which determines the ultimate acceptability or rejection of biomedical devices (ranging from heart valves, to synthetic blood conduits, and to extracorporeal circuitry such as heart-lung machines and artificial kidneys) [52, 53]. A large amount of work has been in progress on the grafting of active biological polymers (such as the highly sulfonated mucopolysaccharide, heparin, one of the body's natural anticoagulants) to a variety of polymeric and inorganic surfaces [54, 55]. This line of work is expanding to include studies of both the grafting to and the spontaneous adsorption to hydrogels of biologically active materials such as heparin and a number of functionally important proteins [56].

Another large area of current work deals with the acceptability of implants in locations of the body where rapid anchoring, infiltration, and ingrowth of biological materials is desired [57]. Such situations include the spontaneous fixation of implantable teeth, the use of porous ceramic and metallic implants in orthopedic locations, the use of reticulated sponge networks for cosmetic surgery and plastic surgery purposes (such as in breast augmentation), and in a variety of other circumstances. Yet another example where interfacial conversion plays a role in the utility of a most common biomedical implant is in providing the proper boundary properties for the sliding surfaces of total hip prostheses, artificial joints and tendons, and muscle connections [58].

10. Biological Phenomena and Interfaces

It is also important to realize that intrauterine contraceptive devices have their interfaces modified during their residence in the uterine cavity by accumulation of spontaneously adsorbed glycoprotein films [59]. The current speculation on their mode of operation is, then, that the adsorbed polymeric films (dominated by a natural protein, from the cervical mucus fluid, which changes its configuration during the course of its adsorption) simply activate the rejection and antibody mechanisms of the adjacent cells. Elaborated antibodies from those cells could be effective in preventing capacitation of sperm which are obliged to swim through that subtly chemically-modified zone. Proper attention to the initial interfacial properties of such implants, even by modifying only the outermost atomic constitution through the application of an interfacially synthesized layer, might significantly improve the acceptance of such devices by eliminating their occasional irritation to the uterine wall and their stimulation of abnormally heavy menstrual flow.

An excellent review of polymers used as surgical implants appeared in 1972 [60], and a more recent monograph reviews the factors influencing blood compatibility of a variety of biomedical polymers [61]. Further, an enormous amount of relevant data on the interfacing of synthetic polymeric materials with biological systems may be found in the technical reports, from contracts funded by such agencies as the National Heart, Lung and Blood Institute. A compilation of over 300 such reports, current through 1975, may be obtained by request to the Biomaterials Program Office, Division of Heart and Vascular Diseases, National Heart, Lung and Blood Institute, Bethesda, Maryland. The reports referenced there are all available through the National Technical Information Service, 5285 Port Royal Road, Springfield, Virginia.

B. Surfaces of Collagen and Keratin

Collagen, the most abundant protein in animal bodies, is a material of increasing use in both technology and medicine. Numerous uses of collagen surfaces in biomedical devices have been described, for example, by Rubin, Stenzel, and co-workers from Cornell Medical College in New York City [62, 63]. All of these uses depend upon the interfacial modification of the collagenous materials, generally accomplished by application of specific catalytic agents (such as enzymes) to remove polymeric fragments (called teleopeptides) which seem responsible for the immunogenic and thrombogenic character of the native material and which cause difficulties when implanted in another living system.

Heterografts of bovine arterial origin have achieved over a decade of excellent surgical acceptance [64]. They, too, have been interfacially modified by digestion with a special enzyme, followed by cross-linking with a dialdehyde starch [65].

Large expanses of biochemical surface which have for too long been ignored by academic interfacial synthetic chemists are those provided by skin and hair. These natural proteinaceous substrates are of obvious commercial, medical and even military interest. Skin biologists, for example, are intensely interested in the surface properties of epidermal and dermal tissues and how those surface properties might be modified to improve the effectiveness of skin as a barrier to entrance of external pollutants and to percutaneous chemical transport of such dangerous environmental agents as the organophosphorous pesticides (now in even greater use since the recent banning of DDT). The pharmaceutical industry is interested, on the other hand, in enhancing the receptivity of skin surfaces to cosmetics, ointments, and medications and in providing interfacial conversion layers that might speed the healing of wounds, provide constant sources of vitamins (such as vitamin C for ulcerated tissues) or deliver long-time-release antibiotic or hormonal preparations. A subject not neglected but very seldom reported upon is the basic surface chemistry of hair, an understanding of which, even in a rudimentary form, currently provides a multimillion-dollar market for bleaches, dyes, colorants, conditioners, permanent waves, and other hairdressing chemicals. A vast opportunity exists in that field alone from a better understanding of interfacial syntheses that might be accomplished at hair surfaces.

10. Biological Phenomena and Interfaces

C. Surgical and Dental Adhesives

Interfacially polymerized cyanoacrylates are currently in use as substitutes for sutures and stitches in general and oral surgery. This work has been pioneered mainly by the U.S. military services and has been most extensively reported by the group of Fred Leonard, then at Walter Reed Army Hospital [66, 67]. Although such work has received much attention in the popular and technical press, an example of substantially greater commercial importance is the use of cyanoacrylates, moisture-curing polyurethanes, and UV-initiated interfacially polymerized agents as dental sealants [68]. These are materials designed to be applied, during a short dental office visit, as a prophylactic treatment to the chewing surfaces of freshly-erupted teeth in children or healthy teeth in adults for the purpose of sealing the pits and fissures where oral bacteria normally reside, propagate, and generate organic acids that encourage caries formation.

D. Maritime Interfaces

Interfacial chemists and physical chemists are now attempting to define the relationships among spontaneous adsorption, polymerization, and cross-linking of molecules at a variety of maritime interfaces [69]. For instance, seawater-setting barnacle cement is exuded and hardened at sites of attachment underwater. Fundamentals of polymer adsorption (including already-formed proteins, polysaccharides, and other biological macromolecules) at these interfaces have yet to be elucidated.

Protein and polypeptide films on aqueous substrates have often been created and characterized as potential models for biological membranes or as specific semipermeable membranes for applications in artificial kidney devices, heart-lung machines, reverse osmosis installations, and desalinization units. Recognizing the phenomenal selectivity of the air-water interface for protein and other organic materials, it is not surprising that bubble scavenging of dissolved organics from sea water might also be of great significance on an oceanic scale. In fact, such scavenging has recently been identified as one of the most significant factors influencing the composition of the oceanic surface and the sea-to-air transferred aerosol [70, 71]. During this transfer—generally by bubble-breaking processes leading to aerosolization of the fragmented bubble cap and of small jet droplets—substantial ion fractionation and organic fractionation occurs across the air-sea interface [72]. It is apparently the long ignored, or unknown, thin surface film of adsorbed glycoprotein and proteoglycan components at the air-sea interface which is responsible for this geophysically important phenomenon [73].

IV. SUMMARY AND PROSPECTS

Before any real understanding of the function or reactivity of biological membranes can be made, it is clear that at least a partial understanding of interfacial polymer structure is necessary. The ability of an enzyme to function in a liposome [74] or of an organism to replicate itself both depend on the physical state of related lipids [23]. At this time, extensive efforts are underway to isolate membrane proteins, glycoproteins, and glycolipids [75] from a wide variety of membranes. Nevertheless, the constituents we know least about are the carbohydrate portions of glycoproteins and glycolipids.

There can be little doubt that glycosubstances play a primary role in both cell recognition and adhesion, as they are usually located in the plasma membrane. As might be expected, the heterosaccharides which hold the key to all this are extremely complex, and identification of the components that play a role in the adhesion of a given membrane has only just begun. The situation is not improved in that these components vary from one cell type to another and that several different types of interaction forces may be involved. The rewards awaiting an improved understanding are great. Not only can we learn how cellular growth, segregation, and replication are controlled but also, hopefully, gain insight into how the outermost layer of a cell is affected during viral infection or malignancy. The key to such understanding is almost certainly the precise arrangement of proteins, lipids, and carbohydrates at biological interfaces.

The examples given here of biochemical phenomena at interfaces are not beyond the ken of polymer chemists attempting the interfacial synthesis of unique, or uniquely-configured, macromolecules. In biochemical situations, however, a specific biopolymer usually automatically provides interface modification of specific substrates. The acquisition of proteinaceous "conditioning" films prior to adhesion of cellular elements in a variety of circumstances has only recently been recognized as an essentially universal event. In the mouth, for example, almost all solid surfaces, be they freshly-erupted teeth or freshly-emplaced oral prosthetic devices, immediately acquire from the rich mixture of surface-active agents provided by the saliva, a marvelously specific glycoprotein film [76]. This film provides an appropriate interfacial layer for subsequent colonization by the oral flora. Similarly, in a much more dilute saline environment—dilute from the viewpoint of containing adsorbable polymeric agents—it has recently been found that spontaneously adsorbed "primary" films of glycoproteins precede biological fouling in sea water. This includes such phenomena as barnacle adhesion to ship bottoms, algae adhesion to engineering and structural materials in the sea, and mussel adhesion to tether lines of buoys and within pipe orifices [77].

A large potential exists for favorable and beneficial interface modification of solid, gel, membranous, and liquid surfaces which have biological, medical, and commercial utility. Interfacial synthesis of polymers, especially including interface conversion of various solids by in situ polymerization at their surfaces, has not yet been properly introduced to the biochemical or biomedical arena. The skills of the "interfacial synthesis community" are needed both for controlling the characteristics of chemical reactions at the numerous surfaces of biological importance and for elucidating the basic physics and chemistry governing those characteristics.

The fundamental salient point made in this chapter is that most biological reactions do, in fact, differ in degree and kind from the phenomena deduced by solution chemists and biochemists. There will continue to be numerous analogies between the traditional chemists' present and future approaches to studying interfacial reactions and syntheses and the needs of the biochemical community. Despite this optimistic forecast, it will be wise to consult Weiss' excellent monograph on the cell periphery [78] before attempting wholesale extrapolation of specific interfacial chemistries into the domain of living cells. The situation is, of course, much more complex than could be properly described in this short overview.

ACKNOWLEDGMENT

D. A. C. acknowledges financial assistance of the Heart and Lung Institute through HL12760-08.

REFERENCES

1. *Epidermal Wound Healing* (H. I. Maibach and D. T. Rovee, eds.), Year Book, Chicago, 1972.
2. *Adhesion in Biological Systems* (R. S. Manly, ed.), Academic, New York, 1970.
3. *Management of Arterial Occlusive Disease* (W. A. Dale, ed.), Year Book, Chicago, 1971.
4. *Surface Chemistry of Biological Systems* (M. Blank, ed.), Plenum, New York, 1970.
5. *Surface Chemistry and Dental Integuments* (A. Lasslo and R. P. Quintana, eds.), Charles C Thomas, Springfield, Ill., 1973.
6. *Chemistry of the Cell Interface* (H. D. Brown, ed.), Academic, New York, 1971.
7. F. R. Eirich, in *Interface Conversion for Polymer Coatings* (P. Weiss and G. D. Cheever, eds.), Elsevier, New York, 1968.
8. Applied chemistry at protein interfaces, in *Advances in Chemistry* (R. E. Baier, ed.), Vol. 145, American Chemical Society, Washington, D.C., 1975.
9. C. E. K. Mees, *The Theory of the Photographic Process*, Ed. 3, Macmillan, New York, 1959.

10. Biological Phenomena and Interfaces 277

10. *Biological Membrane Abstracts,* Informational Retrieval, London, 1973-1976.
11. R. W. Hendler, *Phys. Rev.,* 51:66 (1971).
12. S. J. Singer and G. L. Nicolson, *Science,* 175:720 (1972).
13. See selected reviews in *Recent Progress in Surface Science* (J. F. Danielli, Riddiford, and M. R. Rosenberg, eds.), Academic, London, 1960-1970; also *Progress in Surface and Membrane Science* (J. F. Danielli and D. A. Cadenhead, eds.), Vols. 4-12, 1971-1977.
14. E. Gorter and F. Grendel, *J. Exp. Med.,* 41:439 (1925).
15. J. F. Danielli and H. J. Davson, *J. Cell. Comp. Physiol.,* 5:495 (1934).
16. J. F. Danielli, *Cold Spring Harbor Symp.,* 6:190 (1935); also W. D. Stein and J. F. Danielli, *Disc. Faraday Soc.,* 21:238 (1956).
17. J. F. Danielli, *J. Theor. Biol.,* 12:439 (1966).
18. J. D. Robertson, *J. Biophys. Biochem. Cytol.,* 19:201 (1963).
19. R. S. Criddle and J. Willemot, in *Protides of the Biological Fluids* (H. Peters, ed.), Vol. 15, Elsevier, Amsterdam, 1968, pp. 55-57.
20. D. E. Green, N. F. Haard, G. Lenaz, and H. I. Silman, *Proc. Nat. Acad. Sci. U. S.,* 60:277 (1968).
21. H. M. McConnell, P. Devaux, and C. Scandella, in *Membrane Research* (C. F. Fox, ed.), Academic, New York and London, 1972, pp. 27-37.
22. A. E. Blaurock, *Chem. Phys. Lipids,* 8:285 (1972).
23. J. M. Steim, M. E. Tourtellotte, J. C. Reinert, R. N. McElhaney, and R. L. Rader, *Proc. Nat. Acad. Sci. U. S.,* 63:104 (1969).
24. *Membrane Structure and Its Biological Applications* (D. E. Green, ed.), New York Academy Sci.; published as *Ann. N.Y. Acad. Sci.,* 195 (1972).
25. *First I.C.N.-U.C.L.A., Symposium on Molecular Biology, Membrane Research* (C. F. Fox, ed.), Academic, New York and London, 1972.
26. H. M. McConnell and B. G. McFarland, *Quart. Rev. Biophys.,* 3:91 (1970); also B. G. McFarland, *Chem. Phys. Lipids,* 8:303 (1972).
27. W. L. Hubbell and H. M. McConnell, *Proc. Nat. Acad. Sci. U. S.,* 68:1274 (1971).
28. G. Rouser, G. J. Nelson, S. Fleisher, and G. Simon, in *Biological Membranes: Physical Fact and Function* (D. Chapmann, ed.), Academic, New York, 1968, pp. 5-69.

29. D. A. Cadenhead, R. J. Demchak, and F. Muller-Landau, *Ann. N. Y. Acad. Sci., 195*:218 (1972).
30. D. A. Cadenhead and F. Muller-Landau, *Biochem. Biophys, Acta., 443*:10 (1976); also *Proceedings of the XXIst Colloquim, Protides of the Biological Fluids,* Pergamon, London, 1973, pp. 175-182.
31. D. W. Urry, *Ann. N. Y. Acad. Sci., 195*:108 (1972).
32. I. R. Miller, in *Progress in Surface and Membrane Sci.* (J. F. Danielli, M. D. Rosenberg, and D. A. Cadenhead, eds.), Vol. 4, Academic, New York, 1971, pp. 299-350.
33. B. R. Malcolm, in *Progress in Surface and Membrane Sci.* (J. F. Danielli, M. D. Rosenberg, and D. A. Cadenhead, eds.), Vol. 7, Academic, New York, 1973, pp. 183-229.
34. H. K. Kimelberg and D. Papahadjopoulos, *Biochem. Biophys. Acta, 223*:805 (1971).
35. D. A. Cadenhead, *Ind. Eng. Chem., 61*:22 (1969).
36. D. Branton, *Proc. Nat. Acad. Sci. U. S., 55*:1048 (1966).
37. M. S. Bretscher, *Nature, 231*:225 (1971); *J. Mol. Biol., 59*:351 (1971).
38. H. Hauser, M. C. Phillips, and M. Stubbs, *Nature, 239*:342 (1972).
39. D. Papahadjopoulos, *Biochem. Biophys. Acta, 241*:254 (1971).
40. D. J. Triggle in *Progress in Surface and Membrane Science* (J. F. Danielli, M. D. Rosenberg, and D. A. Cadenhead, eds.), Vol. 5, Academic, New York, 1972, pp. 267-331.
41. P. Overath, H. V. Shairer, and W. Stoffel, *Proc. Nat. Acad. Sci. U. S., 67*:606 (1970).
42. G. Wilson and C. F. Fox, *J. Mol. Biol., 55*:49 (1971).
43. J. K. Raison, J. M. Lyons, R. H. Mehlhoru, and A. D. Keith, *J. Biol. Chem., 246*:4036 (1971).
44. M. Esfahani, A. R. Limbrick, S. Knutton, T. Oka, and S. J. Wakil, *Proc. Nat. Acad. Sci. U. S., 68*:3180 (1971).
45. R. D. Mavis and P. R. Vagelos, *J. Biol. Chem., 247*:652 (1972).
46. H. K. Kimelberg and D. Papahadjopoulos, *Biochem. Biophys. Acta, 282*:277 (1972).
47. S. Wu and J. R. Huntsberger, *J. Colloid Interfac. Sci., 29*:138 (1969).
48. M. Sumper and H. Trauble, *F.E.B.S. Letters, 30*:29 (1973).
49. M. S. Bretscher, *Science, 181*:622 (1973).
50. H. Trauble and P. Overath, *Biochem. Biophys. Acta, 307*:491 (1973).
51. I. Langmuir, *J. Chem. Phys., 1*:756 (1933).

10. Biological Phenomena and Interfaces

52. Problems in evaluationg the blood compatibility of biomaterials, *Bull. N. Y. Acad. Med.*, *48*(2) (1972).
53. Conference on moving blood, *Fed. Proc.*, *30*(5) (1971).
54. G. Grode, S. Anderson, H. Grotta, and R. Falb, *Trans. Amer. Soc. Artif. Intern. Organs,* *15*:1 (1969).
55. V. Gott, J. Whiffen, D. Koepke, R. Daggett, W. Boake, and W. Yound, *Trans. Amer. Soc. Artif. Intern. Organs,* *10*:213 (1964).
56. T. A. Horbett and A. S. Hoffman, Interactions of plasma proteins with radiation-grafted hydrogels, in *Advances in Chemistry* (R. E. Baier, ed.), American Chemical Society, Washington, D.C., 1975.
57. S. F. Hulbert, J. J. Klawitter, B. W. Sauer, and D. M. Bartles, *Characterization of Tissue Ingrowth Into Porous Bioceramics,* Technical Reports on Project NR 032-529, Clemson University, 1973.
58. *Biomaterials for Skeletal and Cardiovascular Applications* (C. Homsy and C. Armeniades, eds.), Wiley, New York, 1972.
59. R. E. Baier and J. Lippes, in *Advances in Chemistry,* (R. E. Baier, ed.), Vol. 145, American Chemical Society, Washington, D.C., 1975.
60. R. I. Leininger, *CRC Crit. Rev. Bioeng.*, *1*:333 (1972).
61. S. D. Bruck, *Blood Compatible Synthetic Polymers,* Charles C Thomas, Springfield, Ill., 1974.
62. A. L. Rubin, R. R. Riggio, R. L. Nachman, G. H. Schwartz, T. Miyata, and K. H. Stenzel, *Trans. Amer. Soc. Artif. Int. Organs,* *14*:169 (1968).
63. K. H. Stenzel, M. W. Dunn, and A. L. Rubin, *Science,* *164*:1282 (1969).
64. N. Rosenberg, A. Martinez, P. N. Sawyer, S. A. Wesolowski, R. W. Postlethwait, and M. L. Dillon, *Ann. Surg.,* *164*:247 (1966).
65. W. A. Dale and M. R. Lewis, *Ann. Surg.,* *169*:927 (1969).
66. F. Leonard, J. A. Collins, and H. J. Porter, *J. Appl. Polym. Sci.,* *10*:1617 (1966).
67. F. Leonard, J. W. Hodge, Jr., S. Houston, and D. K. Ousterhout, *J. Biomed. Mater. Res.,* *2*:173 (1968).
68. M. G. Buonocore, in *Adhesion in Biological Systems* (R. S. Manly, ed.), Academic, New York, 1970.
69. *Proceedings: Third International Congress on Marine Corrosion and Fouling,* Northwestern University Press, Evanston, Ill., 1973.
70. D. C. Blanchard, The borderland of burning bubbles, *Saturday Review,* January 1, 1972, pp. 60-63.
71. Symposium on sea-air chemistry, *J. Geophys. Res., Oceans Atmos.,* *77*(27) (1972).

72. R. E. Baier, *J. Geophys. Res.*, 77:5062 (1972).
73. R. E. Baier, D. W. Goupil, S. Perlmutter, and R. King, *Recherches Atmospheriques, 8*:572-600 (1974).
74. H. K. Kimelberg and D. Papahadjopoulos, *Biochem. Biophys. Acta, 282*:277 (1972).
75. R. B. Kemp, G. M. W. Cook, and C. W. Lloyd, in *Progress in Surface and Membrane Science* (J. F. Danielli, M. D. Rosenberg and D. A. Cadenhead, eds.), Vol. 5, Academic, New York, 1973.
76. R. E. Baier, in *Surface Chemistry and Dental Integuments* (A. Lasslo and R. P. Quintana, eds.), Charles C Thomas, Springfield, Ill., 1973.
77. D. Goupil, V. A. DePalma, and R. E. Baier, Prospects for nontoxic fouling resistant paints, *Proc. Marine Tech. Soc., 9th Annual Conf.*, Washington, D.C., 1973.
78. L. Weiss, The cell periphery, metastasis and other contact phenomena, in *Frontiers of Biology* (A. Neuberger, E. L. Tatus, eds.), Vol. 7, North-Holland, Amsterdam, and Wiley, New York, 1967.
79. *Nobel Symposium No. 34. The Structure of Biological Membranes*, Plenum, New York, 1977.

AUTHOR INDEX

Numbers in brackets are reference numbers and indicate that an author's work is referred to although his name is not cited in the text. Underlined numbers give the page on which the complete reference is listed.

A

Abramson, A. A., 186, 187, 192, 202[33,38]
Acharya, S. P., 64, 74[32]
Adam, N. K., 126, 138[50], 187, 188, 202[36]
Adamson, A. W., 104, 125, 130, 135, 136[1], 138[44,60], 139[66]
Akie, K., 199, 203[52]
Akutin, M. S., 87, 100[49]
Albert, A., 198, 202[42]
Alexander, A. E., 106-108, 110-112, 123, 127, 129-131, 136[7], 137[16,19,26,35, 40], 138[55,58,61], 206, 214, 223, 250[9], 251[42]
Alievskii, P. A., 89, 101[56]
Altier, M. W., 218, 243, 246, 247, 250[26], 252[71]
Amis, E. S., 84, 100[41]
Anderson, J. R., 115, 137[30]
Anderson, S., 270, 279[54]
Armeniades, C., 270, 279[58]
Arth, G. E., 60, 74[17]
Astakhova, A. S., 93, 102[75]
Astaria, J., 184, 202[28]
Auer, E. E., 218, 251[28,29]
Augustine, R. L., 58, 73[4]

B

Baier, R. E., 256, 271, 274, 275, 276[8], 279[59], 280[72,73, 76,77]
Bailey, W. Y., 199, 203[49]
Banba, T., 194, 203[47a]
Banerjee, D. K., 61, 74[26]
Barrett, K. E., 219, 251[32]
Barthalome, E., 222, 225, 251[37]
Barthelemy, M., 65, 75[35]
Bartles, D. M., 270, 279[57]
Bawn, C. E. H., 247, 253[78]
Beaman, R. G., 89, 101[55]
Bekhli, E. G., 85, 100[43,44]
Bell, R. P., 59, 73[8]
Bellamy, W. D., 113, 137[28]
Belyakov, V. K., 192, 198, 202[39c]
Bender, M. L., 83, 84, 100[29,31]
Benson, S. W., 230, 252[59a]
Berg, J. C., 80, 99[8]
Berntsson, P., 72, 75[57]
Bessiere-Chretien, Y., 65, 75[35]
Beste, L., 176, 177, 201[18]
Bevilaqua, B. B., 218, 251[28,29]
Bevington, J. C., 232, 252[63]
Beyler, R. E., 60, 74[17]
Bezdek, M., 218, 250[14a]
Bialecka, E., 70, 72, 75[44,54]
Billmeyer, F. W., 225, 226, 248, 252[51]

Blanchard, D. C., 274, 279[70]
Blank, M., 255, 276[4]
Blaurock, A. E., 264, 277[22]
Bloomfield, J. J., 57, 73[1]
Boake, W., 270, 279[55]
Bose, A. N., 84, 100[35]
Bovey, F. A., 206, 250[15]
Bowden, K., 60, 74[16]
Bradbury, J. H., 78, 81, 87-89, 92, 93, 95-97, 99[3], 101[51,71]
Bradford, E. B., 232, 252[61]
Brandstrom, A., 72, 75[57], 76[58-61]
Branton, D., 267, 268, 278[36]
Brass, P. D., 247, 253[80]
Bretcher, M. S., 267-269, 278 [37,49]
Bretshneider, S., 174, 185, 201[16]
Brientenbach, J. W., 226, 252[52]
Bristow, G. M., 247, 253[76]
Brodnyan, J. G., 218, 250[21]
Brodskii, A. M., 192, 202[39]
Brooks, J. H., 129, 138[57,58]
Brown, D. A., 83, 84, 100[30, 37]
Brown, G. L., 218, 250[21]
Brown, H. C., 60, 62, 64, 74[22,28,31,32]
Brown, H. D., 256, 276[6]
Bruck, S. D., 271, 279[61]
Bruse, W. F., 61, 74[23]
Bunton, C. A., 87, 100[47]
Buonocore, M. G., 273, 279[68]
Burnett, G. M., 222, 225, 251 [35,48]
Byevich, Yu. A., 186, 202[30]

C

Cadenhead, D. A., 257, 266, 267, 277[13], 278[29, 30,35]
Calderbank, P. H., 41, 51, 55, 56[5,16,20]
Carlsson, S., 72, 75[57]
Carothers, W. H., 118, 137[33]
Carraher, C. E., 95, 102[82]
Cary, A., 126, 138[49]

Casogrande, C., 135, 139[70]
Chapman, F. S., 12, 16, 36[3]
Cheer, C. J., 64, 74[36]
Chen, M. C., 83, 100[31]
Chentsova, N. M., 159-161, 165 [9,10]
Chernomordik, R. B., 89, 101[57]
Chielline, E., 198, 202[44]
Chirkov, N. M., 83, 84, 100[26, 39]
Cockbain, E. G., 125, 138[43]
Collins, J. A., 273, 279[66]
Collins, J. C., 60, 74[19]
Conant, J. R., 60, 74[12]
Cook, G. M. W., 274, 280[75]
Cornforth, J. W., 60, 74[18]
Cornforth, R. H., 60, 74[18]
Corrin, M. L., 249, 253[83]
Costich, E. W., 12, 25, 26, 36[5], 55, 56[7]
Coulson, J. M., 107, 136[10]
Crank, J., 82, 99[17], 106, 125, 136[6], 138[46]
Crawford, P. J., 78, 81, 87-89, 92, 93, 95-97, 99[3], 101 [51,71], 102[85]
Criddle, R. S., 261, 277[19]
Crooks, E. H., 217, 250[19]

D

Daggett, R., 270, 279[55]
Dale, W. A., 255, 272, 276[3], 279[65]
Danckverts, P. V., 186, 202[32]
Danielli, J. F., 104, 136[4], 257, 259-261, 277[13,15-17]
Dankwerts, P. V., 82, 99[18]
Davies, J. T., 79-81, 86, 99[6,7, 12], 104, 112, 122, 123, 125, 127, 132, 136[3,4], 137[24,37,41], 138[47,54,65]
Davson, H. J., 259, 277[15]
deBoer, J. H., 105, 136[5]
Defay, R., 108, 109, 111, 137[15]
Dejak, B., 92, 101[70]
Demchak, R. J., 266, 278[29]
Dems, A., 90, 91, 101[63,66-68]
De Palma, V. A., 275, 280[77]
De Puy, C. H., 61, 74[24]
Devaux, P., 262, 277[21]

Author Index

Dillon, M. L., 272, 279[64]
Djurhuus, A., 72, 75[57]
Dubault, A., 135, 139[70]
Duck, E. W., 206, 240, 250[11]
Dunn, M. W., 272, 279[63]

E

Eguchi, W., 222, 251[40]
Eirich, F. R., 256, 276[7]
Elder, M. E., 218, 250[24]
Eliassen, J. D., 234, 252[64]
Elliott, J. J., 83, 100[27]
Enikolopyan, N. S., 90, 101[61]
Entelis, S. G., 83-85, 100[26, 39,40,42-44]
Epel'baum, I. V., 84, 100[40]
Esfahani, M., 268, 278[44]
Everett, H. J., 12, 25, 26, 36 [5], 55, 56[7]
Ewart, R. H., 206, 249[1]
Eyring, H., 119, 137[34]

F

Fainberg, E. Z., 93, 95, 102[72]
Falb, R., 270, 279[54]
Fargo, J., 89, 101[59]
Fedorynski, M., 72, 75[56]
Fedotova, M. I., 162, 165[14]
Fedotova, O. G., 83, 90, 100[25]
Feeks, R. H., 55, 56[4]
Fieser, L. F., 60, 62, 74[13, 14,27]
Fieser, M., 62, 74[27]
Fink, W., 225, 251[44]
Fitch, R. M., 236, 239-241, 252[68]
Fleisher, S., 266, 277[28]
Flory, P. J., 243, 252[74]
Fordyce, D. B., 217, 250[19]
Fox, C. F., 268, 278[42]
Franchini, F. F., 198, 202[44]
Frank, F. J., 60, 74[19]
Franz-Kamenetskii, D. A., 170, 195, 201[3]
Freeman, R. F. S., 247, 253[78]
French, D. M., 223, 251[43]
Fritze, H., 226, 252[52]

Frunze, T. M., 81, 89, 95, 99[14], 101[54,56], 102[83], 172, 173, 185, 194, 201[11,12,15]
Fuller, N. A., 87, 100[47]

G

Gaines, G. L., 104, 113, 136[2], 137[28]
Gardon, J. L., 206, 210, 213-215, 219-222, 225, 229-232, 234- 237, 239, 243, 244, 246-249, 250[5,6,7,12], 251[30,33], 252[57,72], 253[79]
Garg, C. P., 60, 62, 74[22,28]
Gee, G. C., 131, 138[64]
Gerrens, H., 206, 218, 222, 225, 232, 234, 242, 250[13], 251 [27,37,41,44,46]
Gilby, A. R., 130, 131, 138[61]
Glasstone, S., 119, 137[34]
Goldsmith, D. J., 64, 74[33]
Gonzalez, G., 126, 138[48]
Gorter, E., 259, 277[14]
Gott, V., 270, 279[55]
Goupil, D. W., 274, 275, 280[73, 77]
Gouy, G., 112, 137[22]
Grancio, M. R., 249, 253[81]
Grant, G. H., 84, 85, 100[32]
Gray, J. B., 12, 26, 36[6]
Green, D. E., 261, 266, 277[20,24]
Grendel, F., 259, 277[14]
Grisbey, R. G., 217, 235, 250[20]
Grode, G., 270, 279[54]
Grotta, H., 270, 279[54]
Grozdov, A. G., 83, 90, 100[25]
Gustavii, K., 72, 75[57]
Gutsche, C. D., 61, 74[26]

H

Haard, N. F., 261, 277[20]
Hall, H. K., 89, 101[52]
Ham, G. E., 59, 73[5], 168, 200[1]
Hamada, M., 222, 251[40]
Hamann, S. D., 92, 101[69]
Hambly, A. N., 78, 81, 87, 88, 95-97, 99[3]
Harkins, W. D., 135, 139[67], 235, 249, 252[66], 253[83]

Hart, E. J., 218, 251[28,29]
Hartley, G. S., 112, 137[25]
Haselberger, G. S., 80, 99[8]
Hauser, H., 267, 278[38]
Havinga, E., 123, 127, 137[36], 138[53]
Haward, R. N., 206, 249[2]
Hay, P. M., 229, 252[56]
Heathcock, C. H., 65, 67, 75[34]
Heilbron, I. M., 60, 74[16]
Hendler, R. W., 257, 277[11]
Herbeck, R., 222, 225, 251[37]
Herriot, A. W., 73, 76[62]
Herzfeld, S. H., 249, 253[83]
Hess, W. W., 60, 74[19]
Hibberd, G. E., 127, 138[55]
Hine, J. S., 83, 99[22]
Hinshelwood, C. H., 83-85, 100 [28,32-36]
Hodge, J. W., 273, 279[67]
Hodnett, E. M., 93, 102[73]
Hoffman, A. S., 270, 279[56]
Hohnlein, E., 225, 251[44]
Holland, F. A., 12, 16, 36[1]
Holmer, D. A., 93, 102[73]
Hommelen, J. R., 108, 109, 111, 137[15]
Homsy, C., 270, 279[58]
Horbett, T. A., 270, 279[56]
House, H. O., 59-61, 73[6,11], 74[24]
Houston, S., 273, 279[67]
Howard, G. J., 90, 101[62]
Hrabak, F., 218, 250[23]
Hubbell, W. L., 266, 277[27]
Hudson, R. F., 83, 84, 100[30, 37]
Huggins, M. L., 247, 253[77]
Hughes, A. H., 123, 131, 135 137[38]
Hulbert, S. F., 270, 279[57]
Hummel, D., 225, 230, 248, 251[45], 252[58]
Huntsberger, J. R., 268, 278 [47]
Hynkova, V., 218, 250[23]

I

Ikari, T., 192, 197, 199, 202[39a]
Ikeda, K., 200, 203[58]

Imahori, K., 113, 137[27]
Imanishi, M., 66, 75[36]
Ingold, C. K., 83, 87, 99[21]
Isser, S. J., 66, 75[39]
Ito, S., 83, 99[24]
Ivonev, V. V. 90, 101[61]
Iwahashi, M., 135, 139[69]

J

Jiori, C., 200, 203[54,55]
Johnson, C. A., 243, 252[73]
Johnson, W. S., 61, 62, 74[25-27]
Jonczyk, A., 72, 75[55]
Jones, E. R. H., 60, 74[16]
Jones, J. M., 83, 84, 100[29]
Joshi, G. C., 70, 75[45]
Jovanovic, S., 232, 252[62]
Junggren, U., 72, 75[57], 76[58-61]
Jura, G., 117, 137[31]

K

Kaizerman, S., 243, 246, 247, 252[71]
Kamaliddin, A. R., 247, 253[78]
Katz, M., 93, 102[74]
Keith, A. D., 268, 278[43]
Kemp, R. B., 274, 280[75]
Kharit, Y. A., 89, 101[57,58]
Kimelberg, H. K., 267, 268, 274, 278[34,46], 280[74]
King, R., 274, 280[73]
Kinstler, R. C., 87, 100[48]
Kiperman, S. L., 180, 201[23]
Kirmse, W., 69, 75[42]
Kishinevskii, M-Kh., 186, 202[34]
Kivotsukurei, T., 194, 203[47a]
Kiyotsukuri, T., 200, 203[53]
Klawitter, J. J., 270, 279[57]
Klevens, H. B., 218, 224, 250[25]
Knutton, S., 268, 278[44]
Kodani, R. G., 89, 101[53]
Koepke, D., 270, 279[55]
Kögl, F., 127, 138[53]
Kohima, H., 222, 251[40]
Kohnlein, E., 232, 251[46]
Kolesnikov, G. S., 83, 90, 100 [25], 172, 200, 201[4-6,8, 9], 203[56]
Koller, C. R., 89, 101[55]

Kolthoff, I. M., 206, 216, 250 [15,18]
Kondrateva, G. P., 83, 84, 100 [26,39]
Kornblum, N., 58, 73[2]
Kornienko, T. S., 186, 202[34]
Korshak, V. V., 81, 89-91, 95, 99[14], 101[54,56,61,64-66], 102[83], 150, 165 [6a], 168, 172, 173, 183, 185, 194, 199, 200[1c], 201[1d,11,12,15]
Kosabuckii, V. A., 192, 198, 202[39c]
Kozlov, L. V., 86, 95, 100[45], 102[83], 172, 173, 183, 185, 194, 201[11,12,15], 202[26]
Krim, H., 200, 203[53a]
Kruglova, F. L., 95, 102[84], 173, 176, 185, 201[10]
Kuchanov, S. I., 168, 199, 200 [1a,1c], 203[57a]
Kuchner, K., 226, 252[21,21b]
Kudim, T. V., 86, 93, 100[45], 102[78], 141, 165[1]
Kudryavtseva, G. N., 181, 202 [12]
Kulakova, D. G., 89, 101[57,58]
Kurashev, V. V., 81, 89, 99 [14], 101[54,56]
Kuritsyn, L. B., 151, 165[7]
Kwolek, S. L., 81, 86-89, 99 [13], 100[50], 131, 138 [63]

L

Laidler, K. J., 119, 137[34]
Lamm, B., 72, 75[57]
Landfield, H., 225, 251[50]
Langmiur, I., 106, 108, 109, 114, 136[8], 137[18,29], 269, 278[57]
Lasslo, A., 256, 276[5]
Lazocki, Z., 92, 101[70]
Lebedeva, A. S., 89, 101[54]
Lehrle, R. S., 222, 225, 251 [35,48]
Leiniger, R. I., 271, 279[60]
Lenaz, G., 261, 277[20]
Lenz, G., 200, 203[53a]

Leonard, F., 273, 279[66,67]
Levich, V. G., 107, 136[9], 186, 192, 202[29,39]
Lewis, J. B., 80, 99[11]
Lewis, M. R., 272, 279[65]
Lewis, W. K., 80, 99[9]
Ley, G. J., 225, 230, 248, 251 [45], 252[58]
Light, J. C., 229, 252[56]
Limbrick, A. R., 268, 278[44]
Lippes, J., 271, 279[59]
Litt, M., 225, 251[49]
Litvinenko, L. I., 159-161, 165 [9,10,12]
Liu, K. T., 60, 62, 74[22]
Llopis, J., 132, 138[65]
Lloyd, C. W., 274, 280[75]
Logunova, V. I., 159-161, 165 [9,11]
Ludwikow, M., 72, 75[54]
Luisi, P. L., 198, 202[44]
Lurie, A., 58, 73[2]
Lyons, J. M., 268, 278[43]

M

MacRitchie, F., 78, 81, 97, 98, 99[2,15], 106-108, 110-112, 125-127, 129, 130, 136[7, 11,13], 137[19,21,26], 138 [45,48,51,52,59]
Madden, A. J., 59, 74[9]
Magat, E. E., 89, 101[55]
Maibach, H. I., 255, 276[1]
Maiboroda, V. T., 161, 165[13]
Maisuradze, N. A., 90, 101[64]
Makosza, M., 69-72, 75[43,44, 48-56]
Malcolm, B. R., 267, 278[33]
Malofeyevskaya, V. F., 235, 252 [67]
Manly, R. S., 255, 276[2]
Mansoori, G. A., 59, 74[9]
Manyasek, Z., 225, 248, 251[47]
Marchessault, R. H., 246, 253[75]
Mark, P. C., 206, 214, 215, 250 [8]
Marker, L., 229, 252[56]
Markova, G. D., 168, 199, 200[1c]
Maron, S. H., 218, 250[24]
Marsden, J., 123, 137[39]
Martin, D. G., 61, 74[25]

Martinez, A., 272, 279[64]
Mason, S. F., 83, 100[27]
Matheson, M. S., 218, 251[28, 29]
Mavis, R. D., 268, 278[45]
Mayers, G. R. A., 80, 99[10, 12]
McConnell, H. M., 262, 266, 277[21,26,27]
McElhaney, R. N., 264, 274, 277[23]
McFarland, B. C., 266, 277[26]
McKillop, A., 58, 73[3]
McLaughlin, C. M., 41, 56[3]
McMurry, J. E., 66, 75[38-40]
Medalia, A. I., 206, 250[15]
Medved, S. D., 159-161, 165 [9,10], 177, 201[21]
Mees, C. E. K., 256, 276[9]
Mehlhoru, R. H., 268, 278[43]
Melville, H. W., 232, 252
Menger, F. M., 59, 73[7], 74 [10], 82, 97, 99[20]
Metzner, A. B., 55, 56[7]
Mikhailov, N. V., 93, 95, 102 [72], 161, 165[13]
Mikitaev, A. K., 172, 195, 201 [4-8]
Miller, I. K., 216, 250[18]
Miller, I. R., 135, 139[68], 267, 278[32]
Miller, L. A., 89, 101[53]
Mitaishvili, T. I., 90, 101[65]
Mittelman, R., 123, 124, 131, 133, 135, 138[42]
Miyata, T., 272, 279[62]
Moore, D. E., 210, 250[16,17]
Moo-Young, M. B., 55, 56[6]
Morgan, P. W., 78, 81, 83, 86-89, 95, 99[1,13], 100 [50], 101[55], 131, 138 [62,63], 141, 148, 165 [4], 199, 200, 202[41]
Morita, K., 66, 75[36]
Morris, C. E. M., 223, 251[42]
Morton, M., 218, 225, 243, 246, 247, 250[26], 251[50], 252[71]
Muller-Landau, F., 266, 278 [29,30]

Murphey, W. A., 89, 101[60]
Murray, R. W., 229, 252[56]

N

Nachman, R. L., 272, 279[62]
Nagata, S., 222, 251[40]
Naguchi, S., 66, 75[36]
Napper, D. H., 206, 214, 229, 250[9], 252[55]
Neehan, E. J., 206, 250[15]
Nelson, G. J., 266, 277[28]
Nesterov, D. V., 84, 85, 100 [42-44]
Nevel'skii, E. Y., 84, 100[40]
Nicolson, G. L., 257, 263, 266, 277[12]
Nikolyeva, S. S., 161, 165[13]
Nikonov, V. Z., 93-95, 97, 102 [79,86], 147, 165[6], 168, 169, 172, 174, 178, 179, 181-183, 187, 188, 197, 201[2,13,14], 202[24,25,35]
Nishikawa, Y., 194, 203[47a]
Nomura, M., 222, 251[40]
North, A. M., 230, 252[59a,b]
Novokresshchenov, S. S., 164, 165[15]

O

Ogata, N., 192, 197, 199, 202 [39a,b]
Oka, T., 268, 278[44]
Okamoto, I., 199, 203[49]
Oldshue, J. Y., 12, 16, 26, 36 [4], 51, 56[8]
Orienti, M., 198, 202[44]
Ostrovskii, M. V., 186, 187, 202 [33]
O'Toole, J. T., 201, 215, 249[4]
Otto, R. E., 55, 56[4]
Ousterhout, D. K., 273, 279[67]
Overath, P., 268, 278[41,50]
Overnall, D. W., 225, 251[48]
Owens, N. F., 127, 138[52]

P

Palm, V. A., 198, 203[46]
Palmer, R. C., 123, 124, 131, 133, 135, 138[42]

Pande, L., 70, 75[45]
Paoletti, K. P., 225, 226, 248, 252[51]
Papahadjopoulos, D., 267, 268, 274, 278[34,39,46], 280[74]
Papava, G. S., 90, 101[64]
Parfenov, I. V., 164, 165[15]
Parks, A. G., 223, 251[42]
Parts, A. G., 210, 229, 250[16,17], 252[55]
Pastiga, R., 225, 251[49]
Pavlora, S. A., 81, 99[14]
Peaker, F. W., 225, 251[48]
Pelzbauer, Z., 218, 250[23]
Perlmutter, S., 274, 280[73]
Perry, S. G., 87, 100[47]
Pethica, B. A., 129, 138[57]
Phillips, M. C., 267, 278[38]
Picker, D., 73, 76[62]
Pickles, W. J. T., 84, 100[34]
Pitman, I. H., 87, 100[47]
Ponomarenko, A. T., 90, 101[61]
Poos, G. I., 60, 74[17]
Popjak, G., 60, 74[18]
Popov, A. P., 160, 161, 165[10]
Porter, H. J., 273, 279[66]
Posner, A. M., 108, 115, 137[16,30]
Postlethwait, R. W., 272, 279[64]
Powers, R. W., 209, 252[54]
Preston, Y., 199, 200, 203[48]
Priest, W. J., 239, 252[69]
Prigogine, I., 117, 137[32]

Q

Quayle, O. R., 60, 74[12]
Quintana, R. P., 256, 276[5]

R

Rader, R. L., 264, 274, 277[23]
Raison, J. K., 268, 278[43]
Ramos, H. L., 55, 56[4]
Rasmusson, G., 61, 74[24]

Ratcliffe, B. E., 65, 67, 75[34]
Ray, F. E., 60, 74[15]
Reed, G. A., 230, 252[59b]
Rehfeld, S. H., 218, 250[22]
Reinert, J. C., 264, 274, 277[23]
Rezabek, A., 225, 248, 251[47]
Richardson, J. F., 107, 136[10]
Riddiford, 257, 277[13]
Rideal, E. K., 78-81, 86, 99[6], 123, 126, 127, 131, 135, 137[35,37-39], 138[49,54]
Riehl, A., 60, 74[20,21]
Rieveschl, G., 60, 74[15]
Riggio, R. R., 272, 279[62]
Robb, I. D., 222, 223, 251[39]
Robertson, J. D., 261, 277[18]
Robinson, R. A., 87, 100[46]
Rocek, J., 60, 74[20,21]
Rodger, W. A., 41, 56[5]
Rodivilova, L. A., 87, 100[49]
Roe, C. P., 240, 241, 247, 252[10], 253[80]
Roe, J. W., 112, 137[25]
Roginsky, A., 249, 253[83]
Romatowski, J., 232, 252[62]
Rosenberg, M. R., 257, 277[13]
Rouser, G., 261, 277[28]
Rovee, D. T., 255, 276[1]
Rubin, A. L., 272, 279[62,63]
Rushton, J. H., 12, 16, 25, 26, 36[2-5], 40, 41, 51, 55, 56[5-8], 78, 99[5]
Ruysschaert, J. M., 135, 139[68]

S

Sachs, G. D., 209, 252[54]
Salatiello, P. P., 225, 251[50]
Samsoniya, Sh. A., 172, 201[9]
Samsonova, S. A., 199, 203[50]
Samuelsson, B., 72, 75[57]
Santonicola, A. T., 229, 252[56]
Saraga, L., 117, 137[32]
Sarett, L. H., 60, 74[17]
Satatiello, P. P., 218, 250[26]
Sauer, B. W., 270, 279[57]
Savelova, V. A., 160, 165[12]
Savelova, V. I., 159-161, 165[9,10]
Sawyer, P. N., 272, 279[64]
Sawyer, W. M., 218, 250[22]

Scandella, C., 262, 277[21]
Schaefer, V. J., 106, 108, 136[8]
Schneider, C., 225, 230, 248, 251[45], 252[58]
Schnell, H., 200, 203[53a]
Schulman, J. H., 123, 125, 131, 137[40], 138[43]
Schulz, G. V., 219, 223, 232, 251[31], 252[62]
Schwartz, G. H., 272, 279[62]
Sekine, Y., 200, 203[58]
Serafin, B., 71, 72, 75[48-52,55]
Serzhent, E., 198, 202[42]
Shairer, H. V., 268, 278[41]
Sharikov, Yu. V., 162, 165[14], 168, 169, 201[2]
Shilyakova, G. N., 93-95, 102 [79], 172, 178, 197, 201[14]
Shimomura, Y., 200, 203[53]
Shpital'nyi, A. S., 89, 101 [57,58]
Shpital'nyi, M. A., 89, 101[58]
Silberberg, A., 110, 137[20]
Silman, H. I., 261, 277[20]
Simon, G., 266, 277[28]
Singer, S. J., 257, 263, 266, 277[12]
Singh, N., 70, 75[45]
Smirnova, O. V., 172, 195, 199, 201[4-9], 203[50]
Smith, M. V., 206, 249[1]
Smith, W. V., 221, 222, 225, 231, 232, 251[34,36]
Sokolov, L. B., 78, 93-95, 99 [4], 102[75-81,84,86], 141, 143, 144, 146-150, 152-155, 157, 159-162, 164, 165[1-3,5,6,8-11, 14,15], 168, 169, 172, 173, 176-178, 181, 183, 185, 187, 188, 194, 197, 198, 200, 201[2,10,14, 20,21], 202[24,35], 203 [45,57]
Sokolova, D. F., 160, 161, 165 [11], 198, 203[45]
Solomon, D. H., 92, 101[69]
Sorokin, A. Y., 89, 101[58]

Speck, S. B., 89, 101[60]
Stanley, J. H., 184, 202[27]
Stannett, V., 225, 246, 251[49], 253[75]
Starks, C. M., 68, 70, 75[41]
Steim, J. M., 264, 274, 277[23]
Stein, W. D., 259, 260, 277[16]
Stenzel, K. H., 272, 279[62,63]
Stern, O., 112, 137[23]
Stockmayer, W. H., 206, 213-215, 232, 249[3]
Stoffel, W., 268, 278[41]
Stokes, R. H., 87, 100[46]
Strichman, G. A., 176, 177, 201 [17]
Stubbs, M., 267, 278[38]
Sumper, M., 268, 278[48]
Sundberg, D. C., 234, 252[64]
Sundet, S. A., 89, 101[60]
Suzuki, A., 64, 74[31]
Sweeting, O. J., 229, 252[56]
Swift, J. D., 92, 101[69]
Sykes, P., 198, 203[47]

T

Tabushi, I., 70, 71, 75[46,47]
Takahashi, N., 70, 71, 75[46,47]
Tarasov, A. I., 168, 199, 200[1c]
Tarkowsky, H. L., 232, 252[61]
Tarnowiecki, H., 226, 252[52]
Taylor, E. C., 58, 73[3]
Taylor, R. P., 232, 252[63]
Temkin, M. I., 180, 201[22]
Tent, G. S., 209, 252[54]
TerMinassian-Saraga, L., 107, 118, 125, 129, 136[12], 138[56]
Thomas, H. R., 219, 251[32]
Tiger, R. P., 84, 100[40]
Timofeeva, G. I., 91, 101[66]
Tokarev, V. I., 160, 161, 165[10]
Tommila, E., 84, 97, 100[38]
Tordai, L., 108-110, 136[14], 137[17]
Tourtellotte, M. E., 264, 274, 277[23]
Trauble, H., 268, 278[48,50]
Trebbi, G. F., 217, 250[19]
Treybal, R., 186, 192, 202[31]
Trice, V. G., 41, 56[3]

Author Index

Triggle, D. J., 268, 278[40]
Tsai, C. H., 236, 239-241, 252[68]
Tsutsumi, Y., 83, 99[24]
Tsuyoshi, K., 199, 203[52]
Tumakov, S. G., 192, 198, 202[39c]
Turetski, L. V., 93-95, 102[76,77], 141, 147, 148, 165[1,5]
Turska, E., 90, 91, 101[63,66-68], 196, 202[40]
Tuthill, J. D., 55, 56[4]
Tweet, A. G., 113, 137[28]
Tynzye, V., 176, 177, 201[19]

U

Ugelstad, U., 206, 214, 215, 250[8]
Uhl, V. W., 12, 26, 36[6]
Ulevich, I. N., 218, 250[24]
Urbanski, T., 71, 75[48]
Urry, D. W., 266, 278[31]

V

Vagelos, P. R., 268, 278[45]
Vander Hoff, B. M. E., 206, 222, 223, 225, 230, 240, 249, 250[14], 251[38], 252[60], 253[82]
Vander Hoff, J. W., 206, 207, 232, 242, 250[10], 252[61]
Van Krevelen, D. W., 82, 99[19]
Vanstone, A. E., 66, 75[37]
Vanzo, E., 246, 253[75]
Vasil'ev, A. V., 90, 101[65]
Vasnev, V. A., 90, 101[65], 168, 199, 200[1a,b,c], 201[1d]
Velichkova, R. S., 90, 101[61, 64]
Venkataraman, H. S., 83, 84, 100[28,36]
Veyssie, M., 135, 139[70]
Viallard, A., 81, 99[16]
Viharainen, T., 84, 97, 100[38]

Vinogradova, S. V., 89, 90, 101[54,61,64], 150, 165[6a], 168, 199, 200[1b,c], 201[1d]
Vorobyev, N. K., 151, 165[7]
Vroom, R. A., 82, 99[19]

W

Wakil, S. J., 268, 278[44]
Wall, T. F., 209, 252[54]
Ward, A. F. H., 108-110, 136[14], 137[17]
Warnhoff, E. W., 61, 74[25]
Wasley, W. J., 89, 101[53]
Watson, F. M., 247, 253[76]
Watterson, J. G., 210, 250[16,17]
Wawrzyniewicz, M., 69, 75[43]
Weeden, B. C., 60, 74[16]
Weiss, L., 276, 280[78]
Weitz, H. M., 222, 225, 251[37]
Wepsic, J. C., 229, 252[56]
Wesolowski, S. A., 272, 279[64]
Westheimer, F. H., 62, 64, 74[29]
Whiffen, J., 270, 279[55]
Whitehurst, J. S., 66, 75[37]
Whitfield, R. E., 89, 101[53]
Whitman, W. G., 80, 99[9]
Wiggill, J. P., 80, 99[7], 125, 138[47]
Willeboordse, F., 83, 99[23]
Willemot, J., 261, 277[19]
Williams, D. J., 249, 253[81]
Williams, E. G., 84, 100[33]
Wilson, G., 268, 278[42]
Wittbecker, E. L., 89, 101[55]
Wright, P. V., 90, 101[62]
Wu, S., 268, 278[47]

Y

Yeliseyeva, V. I., 235, 252[67]
Yoshida, Z., 70, 71, 75[46,47]
Yound, W., 270, 279[55]

Z

Zaweski, E. F., 61, 74[24]
Zbinden, R., 198, 202[43]
Zimmit, W. S., 226, 228, 229, 252[53]
Zubov, P. I., 235, 252[67]

SUBJECT INDEX

A

Absorption coefficient, 53
Accessibility factor, 123, 133
Accessibility of reactive
   groups, effect on
   reaction rate, 133
Acid chlorides, 54, 87, 89,
   90, 97, 185, 193, 198
   (see also Diacid
   chlorides)
   activation energies, 84
   fluorinated, reaction with
      amines, 159, 184
   hydrolysis, 86, 96, 160, 163,
      185, 194
   reaction with amines at
      liquid-vapor interfaces,
      142-164
   reaction with amines,
      mechanism, 83, 85
   reaction with aromatic
      amines, 84
Activation energy, 84, 90, 96,
   97, 109, 119, 121, 122,
   125, 131, 147, 150, 151,
   226
Activation entropies, 84, 97
Activity Coefficients, 87
Acyloin reaction, 57
Adamantane, reaction with
   dichlorocarbene, 70, 71
Additives, 33
   salts, 86-98
Adhesives, surgical and
   dental, 273
   cyanoacrylates, as sutures,
      stitches, dental seal-
      ants, 273

Adipoyl chloride, 173, 185
Adsorption, 97, 98, 104-136, 197
   coefficients, 51-53, 180, 187
   energy, 106
   localized, 113
   monomers, 82, 187
   rate of, 108, 128, 238
   sites, availability, 109
   solute, 105-109
Alcohols:
   allylic, conversion to ketones,
      66
   conversion to chlorides, 71
   dehydration of, 66
   distribution at air/water
      interface, 114-117
   oxidation of, 62
   oxidation to ketones, 60
Alkylation(s), 57-58
   active methylene compounds, 73
   carbanion, 73
   carbon, 71
   Darzens condensation, 72
   effect of stirring on, 46
   indene, 72
   ketones, 72
   S-phenylglycolonitrile, 72
Amides, of carboxylic and arylsulfonic acids, mechanism
   of formation, 96-98, 160,
   161
Amine-ended, 97
Amines (see also Amides, Aromatic
   diamines, Diamines)
   reaction with acid chlorides,
      83-85, 89, 159, 160
Anhydrides:
   reaction with diamines, 90
   reaction with diols, 90

Aniline, reaction with benzoyl chloride, 151
Aromatic diamines, reaction with phosgene, 157-159
Autocatalytic, 85

B

Back-diffusion, 105, 108
Benzoyl chlorides, hydrolysis of, 83
  reaction with aniline, 151
Benzoyl-o-toluidine, oxidation of, 59
Bilayer, 260, 266, 267
Biological interfacial phenomena, 255-276 (see also Membranes)
  cell recognition and adhesive, 274
  examples--healing of lacerations, surgical incisions, synthetic blood conduits, bioadhesive events, accumulation of biological deposits, 256
  summary and prospects, 274
Biological polymers and interfaces, 256-276 (see also Membranes)
  potential treatment of diseases as leukemia and enzyme deficiencies, 269
Biomedical implants, importance of interface, 270
  examples--heart valves, grafting, heart-lung machines, kidneys, cosmetic and plastic surgery, hip prostheses, artificial joints and tendons, muscle connections, 270
Biomembrane organization, 257-269
Block structure, 168-200
Borneol, oxidation to camphor, 59
Butyl acrylate, emulsion polymerization of, 215

C

Carbanion alkylations, 69
Carbenes, phase, transfer catalytic reactions of, 70
Catalysis, 68, 73, 104
Cell membrane (see Membranes)
Ceric ion oxidations, 59
Chemical potential(s), 124
Cholesterol formate, rate of hydrolysis, 132, 133
Chromium compounds, uses in oxidations of alcohols to ketones (see Ketones)
Collagen surfaces, 271, 272
Collins reagent, 60
Copacamphen, synthesis of, 66
Copolyamide synthesis, 172, 198
Copolycondensation, 2, 5, 93, 94, 145-149, 168-200
  adsorption of monomers onto reaction surface as limiting stage, 184-187
  aryl dicarbonyl and disulforylchlorides with bisphenols, 5
  2,2-bis(4-aminophenyl)propane and 4,4-dihydroxybinaphthyl-1,1' with terephthalayl chloride, 196
  composition, 175-200
    effects of hydrolysis and hydrogen chloride, 193
    effects of side reaction, 193, 195
    of ethylene diamine and hexanediamine with adipoyl chloride, 189
    excess of intermonomer, 194
    under external kinetic region, 170-180
    under internal kinetic region, 170-176
    under kinetic regions, 175-196
    under other regions, 192, 193
    under transitional regions, 190-192
    where adsorption is limiting, 188

[Copolycondensation]
   where reaction is limited
     by mass transfer, 182-188
   conditions of synthesis,
    197-201
     concentration, 199
     emulsifiers, 200
     monomer pairs, 198
     organic phase, 199
     procedure, 200
     reactivities, 198
     stirring, 199
   of diamines with two or more
     acid chlorides, 184
   diffusion, 195
   effect of hydrolysis on, 185
   of ethylene diamine and
     p-phenylene diamine with
     adipoyl chloride, 183
   mass transfer, 168-200
   mass transfer as limiting
     stage, 181-200
   mechanism, 186
   parameters, 171-174, 179,
     183, 187, 190, 191, 195
   of tetramethylene diamine and
     decamethylene diamine
     with isophthaloyl
     chloride, 178, 179, 182,
     183, 186, 188, 190, 197
Copolymer-composition, 145,
   175-200
   synthesis using liquid-vapor
     systems, 145-148
   types, 169
Cornforth reagent, 60
Cyanocrylates, as sutures,
   stitches, dental sealants, 213
Cyclic products--polyamides,
   145
Cyclohexanone, prepared from
   cyclohexanol, 62
2-Cyclopentene-1,4-dione,
   preparation of, 61
Cyclopropanes, synthesis of,
   69

D

Decalin diols, conversion to
   diketones, 61
Decalones, 62
Dehydration of alcohols, 66
Desorption, 124, 125, 128-136
   measure of rate of, 126, 129
   of products, 82
   reaction with acid chlorides
     at liquid-vapor interface,
     145, 147, 148, 152
   reaction with oxalyl chloride,
     146, 162
   reaction with sebacoyl chloride,
     97
Diacid chlorides, 89, 90, 159,
   162, 168, 175, 192 (see
   also Acid chlorides)
   fluorinated, reaction with
     diamines, 143
   reaction with diamines, 92, 95,
     96, 126, 142, 145, 149,
     152, 177
   reaction with diols, 90, 95
Diamines, 86-88, 90, 92, 93, 95,
   96, 126, 142, 146, 154,
   168, 175, 178, 182-184,
   187-189, 194, 195
β-Dicarbonyl compounds, 58
Dichlorocarbene, formation, 70
   reaction rates with olefins, 70
   reaction with adamantane, 70
Dichlorocyclopropanes, formation of, 69
Dichloromethylphenyl sulfide,
   reaction with styrene, 70
Dichloronorcarane, synthesis of,
   70
Dielectric constant, medium,
   84, 87
Diffusion, 10, 54, 79-98, 105,
   112, 113, 124, 170-200,
   238 (see also Mass Transfer
   and Transport
   coefficient, 80, 106, 107, 113,
     129, 181, 183-185

[Diffusion]
  external, 170
  internal, 170
  rate, 174, 182
Dihydrocholesterol, conversion to cholestanone, 60
Diisocyanates, reaction with diamines, 90, 196
  reaction with diols, 90
Diketones, 62
Dipolar aprotic solvents, 59
Dispersion, 33-36
Dissociation constants, 98
Distribution coefficient, 134, 176, 178, 179, 181
  of diamines, 176, 178
Disulphonyl chlorides, 89
Drag coefficient, 26 (see also Power number)

E

Economic factors, 6, 164
Eddy diffusion, 10, 80
Eddy viscosity, 27
Einstein diffusion equation, 106
Electrical dipole moment, 127, 132
Electrical potential barrier, 111, 114, 122, 123, 124, 125
  effect on rate, 133
Emulsion copolycondensation, 177
Emulsion polymerization, 206-249 (see also Smith-Ewart)
  characteristic parameters, 216
  conversion rate, 208-230, 242
  diffusion coefficient, 237
  experimental tests for model,
  Flory-Huggins constant, 243-246
  general features, 206
  Harkins hypotheses, 235, 236, 240

[Emulsion polymerization]
  initiator parameter, 216-217
  interfacial tension, 243, 245-247, 249
  mechanistic model, 207, 238
  molecular weight, 212, 215, 221, 222, 232, 233
  monomer volume fraction, 215, 243
  Morton-Kaizerman-Altier equation, 244
  nucleation, 223, 238, 240
    critique of theories, 241
  particle radius, 242
  particle size, number and distribution, 241, 242
  Q-function, 213-215, 229, 230
  radical utilization efficiency, 222
  soap parameter, 217-218
  stability, 125
  swelling, 242
    partial, 245
    thermodynamics, 243
  theoretical prediction of size, conversion rate and molecular weight, 208
  validity for model, 230
  volume growth rate constant, 218, 219
Emulsions, 24, 33-36, 54, 63
Energy field, 104
Enolate ion, 58
Epimerization, 60-62
Epoxyolefin cyclization, 64
Equilibrium drop size, 41-55
Equilibrium, effect of mixing on, 13-36, 41
Esterification, 84
Esters, hydrolysis of, 131
Ethylenediamine, reaction with oxalyl chloride, 145, 148
Extraction alkylation, 72, 73
  (see also Phase transfer catalyses)

F

Fick's law of diffusion, 79-98
Flory's principle, 169

Subject Index

Flow, 38-56
Fluid kinetics, 11-36
Fluid mechanics, 12-36, 38, 40-56
Fluid motion, 11-56
Fluorinated acid chlorides, reaction with diamines, 143
Free energy of adsorption, 105
Froude group, 41

G

Gas-liquid reactions, effect of stirring on, 31-36, 50-56 (see also Liquid-vapor)
  mechanism, 93, 94
Glycolipids, 258, 262, 274
Glycoproteins, 258, 261, 267, 271, 274, 275

H

Halocarbene formation, 69
Heat transfer, 39-56
1,6-Hexanediamine, 144, 146, 148, 149, 151, 157, 163, 173, 185
  copolyamide with phosgene and oxalylchloride, 153
High speed stirring, 38-55
Hydraulic regime, 11-36
Hydroboration, 64-66
Hydrodynamic boundary layer theory, 107
Hydrogenation, 58
  of fatty oils, effects of stirring on, 50
Hydrolysis, 84, 86, 89, 90, 96, 145, 147, 149-151, 159-161, 163, 169, 194

I

Ionic strength of medium, effects of, 86-98
Imidazol catalyzed hydrolysis, 59
Immiscible liquids, effect of mixing on, 32-56
Impeller, performance, 11-36
  types, 16-36
  types of flow from, 24-36
  variables, 10-36
Inhibitors, effect on reaction rates, 124
Interbipolycondensation, 168, 171, 196
Interfacial area, effect of stirring on, 39-56
Interfacial potential, 127, 130, 134
Interfacial precipitation, 129
Interfacial pressure, 126, 127, 131
Interfacial pressure barrier, 110, 114, 122-125
Interfacial resistance, 79, 126
Interfacial tension(s), 33, 40-43, 50, 81, 97, 114, 187
Interfacial viscosity, 127, 130
Isoborneol, oxidation to camphor, 59
Isocaranol, conversion to a ketone, 64
Isophthaloyl chloride, reaction with diamines in copolycondensations, 179, 180, 182, 183, 185-187, 189-192
Isophthaloyl difluoride, reaction with aniline, water, 159, 160
3-Isothujopsanone, from alcohol, 64

J

Jones reagent, 60

K

Keratin surfaces, 271-272
Ketone, 60-63, 66
Kinetics, chemical, 3, 11, 39, 78-98, 127-130, 158, 169, 170, 179, 231
  fluid, 11

[Kinetics]
  macroscopic, copolycondensation, 169, 170, 175-193, 197
  salt effects, 87
Kinetic region of copolycondensation, 175, 176
  external kinetic region, 170-176
  internal kinetic region, 170-176

L

Latex particle, 207, 208, 215
Lauric acid, desorption of, 125
Lipid bilayer, 261, 267, 268
Lipids, 258, 261, 262, 264-266, 268
Lipoproteins, 258, 261
Liposomes, 266, 267
Liquid-vapor interfacial condensations, 141-164
  copolycondensation, 93-96

M

Mass flow, 13-36
Mass transfer, 31, 79-98
  coefficient, 51, 53, 186, 192
Mechanisms, 3, 78-98, 118, 124
  amide formation, 160-161
Membranes, biological, 1, 5, 256-276
  bilipid layer, 259, 263, 264
  carbohydrates, 258, 262
  cell, 257, 258
  Danielli-Davson concept, 259, 260
  Langmuir's "duplex" film, 269
  model, 257, 259
  modified, 261
  myelin, 261, 262
  permeability, 260, 267
  proteins, 258-263, 267
ℓ-Menthol, conversion to ketone, 62

Mesitoyl chloride, hydrolysis, 83
Methyl acrylate, emulsion polymerization, 223
2-Methylcyclohexanone, preparation of, 61
Methyl methacrylate, emulsion polymerization, 206-249
Mixing, flow patterns, 15-36
  flow regime, 11-36
  hydraulic regime, 11-36
  mechanical factors, 12-36
  principles, 10-36
Molecular weight, 8, 86, 88, 89, 91, 93, 96, 126, 145, 149, 150, 158, 159, 161-163, 169, 172, 175-178, 191, 198, 199
  liquid, vapor systems, 151-154
  production of narrow distribution, 118

N

Newtonian liquid forces, 40-56
p-Nitrophenyl laurate, hydrolysis of, 59
Nucleic acid, 2
Nucleophilic displacements by phase transfer catalyses, 69

O

Oleic acid, oxidation of monolayers, 128, 135
Organic solvent, effect on reactions, 87
Oxalyl chloride, copolymer with phosgene, 148, 149, 156, 157
  reaction with diamines, 144, 162, 163
  reaction with ethylenediamine, 145, 152, 153
Oxalyl fluoride, reaction with diamines, 144
Oxidation, fats and fatty acids, 131

P

Partition coefficients, amines, 87
  equilibrium, 176
  functions, 119, 120
Permanganate oxidation, 59
Permeability coefficient, 79, 80 (see also Transfer coefficient)
pH effects, 86-98
Phase boundaries, 104-136
Phase ratio, 42, 44
Phase transfer catalyses, 68
Phenolate salts, 58
Phenyl acetonitrile, ethylation of, 71
1,4-Phenylenediamine, reaction with acid chlorides, 144, 151, 177
  reaction with phosgene, 158
Phosgene, copolyamide with oxalylchloride, 146, 148, 149, 156, 157
  reaction with 1,4-phenylenediamine, 151-153, 158, 159
  reactions, 154, 159
Phospholipids, synthetic, 257
Phthaloyl chlorides, kinetics of hydrolysis and esterification, 84
Piperazine, reaction with terephthaloyl chloride, 96-98
Polarity, of phases, 82, 87, 91
Polyamidation, 81, 83, 172
  activation energies, 96, 97
  mechanism, 96-98
  side reactions, 88-90
Polyamide, 144, 194
  branched, 88, 89
  fluorinated, 144
  from liquid-vapor systems, 142, 143
  interfacial precipitation, 126
Polyamidesters, 200
Polyarylate, 91

Poly(bis-chloroformates), 90
Polycarbonate, 91
Poly(decamethylene isophthalamide) film, content, 189
Poly(ethylene silylenes), mechanism of formation, 95
Poly(1,6-hexamethyleneoxamide), 153
Polyorganophosphates, 2
Polyoxamides, 144, 155, 156
Poly(1,4-phenylene oxamide), 155, 156
Polythioesters, 144
Polythioureas, 144
Polyureas, 144
Polyurethane, 195, 196, 273
Polyurethane carbonates, 200
Polyurethane, formation of cyclic, 89
Power number, 26-36
Protein and polypeptide films, 273
  applications in kidney devices, heart-lung machines reverse osmosis, desalinization, 273

R

Rate constants, 85, 90, 97, 123, 131, 132, 134, 135, 159, 209, 213, 216, 226
Rate of adsorption, 108-112, 189
Rate of desorption, 107
Rate of evaporation, 107
Rate of particle nucleation, emulsion systems, 208
Rate of radical absorption, emulsion systems, 208-249
Reaction rate(s), 54, 105, 113-136, 176
  effect of mixing on, 11-36
  interface, 119-136
  order, 90, 95-97
Reagent ratio index, 152, 153
Relative reactivity constant, 174-178, 180
Reynolds number, 26-36, 40-56

S

Salt effects, 87-98
Saponifications, 29
Sarett reagent, 60
Sativene, synthesis of, 66
Schotten-Baumann reaction, 1
Sebacoyl chloride, reaction with aliphatic diamines, 97, 153
Side reactions, 88-90, 96
Silanediols, polycondensations of, 92
Smith-Ewart theories, emulsion polymerizations, 206, 249
  deviation, 227-232
  experimental tests for model, 215, 223, 225
  validity, 230, 232
  value, 211, 212
Solution, precipitation polycondensation, 88-98
Solvent effects, 86-98
Steroidal diol, conversion to ketol, 66
Stirring, 4, 10
  effect on copolycondensations, 172-174
  equipment, 11-36
  power, 25-27
  rate, 4, 5
Stopped, flow kinetics, 85
Styrene, emulsion polymerization of, 206-249
Sulfonic acid, 58
Sulphonyl chloride, hydrolysis, 89
Surface effects, 78
Surface tension, 118, 147, 187, 217, 236
  nitrogen-water, 161
Suspension polymerizations, 55

T

Terephthaloyl chloride, reaction with piperazine, 96-98
Terpenoid alcohols, conversion to ketones, 64
Tetrachlorohydroquinone, oxidation, 59
Thermodynamic salt effects, 87
Transfer coefficient, 80-82
Transport, 79, 81-98
Traube's coefficient, 147, 188, 189
Tricyclic alcohols, conversion to ketones, 65
  reduction of, 67
Triolein monolayers, oxidation of, 124, 134
Turbulence, 26-36, 38-56 (*see also* Eddy viscosity)
Two-phase oxidations, 60-73

U

Ureas, 83
Urethane, 83

V

Vinyl acetate, emulsion polymerization of, 223
Vinyl chloride, emulsion polymerization of, 215
Viscosity (of phase(s)), 42, 82, 91

W

Waring blendor, 38
Weber number, 40, 41, 43